The Insect Crisis

The Insect Crisis

The Fall of the Tiny Empires That Run the World

OLIVER MILMAN

W. W. NORTON & COMPANY
Independent Publishers Since 1923

First American Edition 2022

Printed in the United States of America

For information about permission to reproduce selections from this book, write to
Permissions, W. W. Norton & Company, Inc., 500 Fifth Avenue, New York, NY 10110

For information about special discounts for bulk purchases, please contact
W. W. Norton Special Sales at specialsales@wwnorton.com or 800-233-4830

Manufacturing by Lakeside Book Company
Book design by Chris Welch
Production manager: Lauren Abbate

Library of Congress Cataloging-in-Publication Data

Names: Milman, Oliver, author.
Title: The insect crisis : the fall of the tiny empires that run the world / Oliver Milman.
Description: First American edition. | New York, NY : W. W. Norton & Company, 2022. |
Includes bibliographical references and index.
Identifiers: LCCN 2021042932 | ISBN 9781324006596 (hardcover) | ISBN 9781324006602 (epub)
Subjects: LCSH: Insects—Conservation. | Insect populations. | Rare insects.
Classification: LCC QL467.8 .M55 2022 | DDC 639.97/57—dc23
LC record available at https://lccn.loc.gov/2021042932

W. W. Norton & Company, Inc., 500 Fifth Avenue, New York, N.Y. 10110
www.wwnorton.com

W. W. Norton & Company Ltd., 15 Carlisle Street, London W1D 3BS

1 2 3 4 5 6 7 8 9 0

CONTENTS

The Insect Crisis

Prologue

The first inkling of the cataclysm was the deathly stillness. The countryside, suburban gardens, and urban parks, their soundtracks now muffled, became lifeless imitations of themselves. No more rumbling buzzsaw of a passing bee, no metronomic chirping of a cricket, no nagging whine of a famished mosquito.

Landscapes suddenly felt as flat as the oil paintings they inspired, perhaps even less vivid given the riot of colors that had been wrenched from the ecological palette once the iridescent butterflies and flamboyant beetles were gone.

The world's insects had vanished, but the lag of human inertia meant that the first howl of horror, oddly, came not from us but rather from birds. The skies and forests were the settings for increasingly frantic bluebirds, nighthawks, woodpeckers, and sparrows as they searched for aphids, moths, and other meals no longer there. The deficit was huge—around 200,000 insects had to be served up to raise a single swallow chick to adulthood. Now there were none. In all, half of the roughly 10,000 species of birds on Earth starved to extinction, their withered corpses strewn on the ground and within barren nests.

An array of dead bodies—birds, squirrels, hedgehogs, humans, in fact anything that set foot on land and was mortal—began to build up across valleys, hills, parks, and neglected city apartments. Blowflies, which laid maggots able to consume 60 percent of a human corpse within a week, were now absent, as were the moths, dermestid beetles, and the rest of the cavalcade of insects that previously arrived to break down the deceased. Bacteria and fungi were still there to do the job, but at a far slower pace. It wasn't enough. The rotting carcasses and putrid smell triggered public revulsion, until that, too, became normal.

As if the world around us was conspiring to turn our stomachs, the lingering flesh and bone was compounded by a tsunami of feces, seemingly everywhere, left wherever it fell. Farmers in Australia had previously endured a painful lesson on the importance of the right sort of dung beetle being present after cattle were first introduced by European settlers. Now, the continent was awash with vast areas of useless land caked in livestock manure that the native beetles, more used to marsupial dung, weren't able to break down. With 8,000 species of dung beetle—a group that had been doing a thankless cleanup job for the planet for at least 65 million years—wiped out globally, this disaster now repeated itself on a far grander scale, with feces from wildlife and livestock pockmarking the planet unchecked like a foul plague. Millions of acres of land were laid to waste. Felled trees and leaves also started to accumulate, stubbornly refusing to disintegrate back into the earth.

Disgust and then alarm began to take hold around the world. Environmental groups mobilized, holding rallies featuring people dressed as bees, while politicians huddled in emergency meetings and issued hasty promises of action. It felt like something could be done.

Then the food supply disintegrated. More than a third of global food crop production was dependent on pollination from thousands of bee species as well as other creatures, such as butterflies, flies, moths, wasps, and beetles. With pollinators gone, a global conveyor belt of food production shuddered to a halt, with sprawling fields of fruit and vegetables left to wither away. Farmers no longer needed to spray pesticides to vanquish pests but lamented that the invaders would have little to destroy anyway.

Items such as apples, honey, and coffee faded away from supermarkets and became expensive luxuries. The disappearance of cecidomyiid and ceratopogonid midges, the unheralded pollinators of the cacao tree, cut off the supply of chocolate. People openly wailed in the streets at this loss; rates of depression and anxiety soared.

The loss of bees stripped the world of readily available items such as strawberries, plums, peaches, melons, and broccoli, with the remaining fruit and vegetables oddly shaped and pathetically shriveled. Mercifully, an apocalyptic starvation event was averted thanks to our reliance on staples such as wheat, rice, and maize, which are pollinated via the wind.

Still, meals became blander and less nutritious, even in wealthy countries. Without access to fruit, vegetables, nuts, or seeds, millions of people eked out a grim diet based around oats and rice. Any thought of consuming a mango or almond became a decadent fantasy, before the experience faded from collective memory entirely. With no chilies, cardamoms, coriander, or cumin, curries became a historical dish. Restaurants of various hues, struggling to even source tomatoes or onions, closed en masse. Cows, once fed a diet of now-scarce alfalfa, dwindled. Fewer cows meant shortages of milk and dairy, which in turn meant no cheese, yogurt, or ice cream.

Governments started to assemble armies of workers to hand-pollinate crops, although this proved wildly more expensive and far less efficient than the 100-million-year-old codependency that had evolved between insect pollinators and plants. A rash of new companies launched swarms of drones and robotic bees in an attempt to replicate the real thing. These efforts proved insufficient.

As with most calamities, the poor and vulnerable fared worst. More than 800 million people globally were malnourished before the insects vanished, and many of them were pushed over the edge into starvation once the nutrients from pollinated crops receded. Cases of childhood blindness jumped as vitamin A, derived largely from fruit and vegetables in the developing world, was eliminated from diets. The curses of malaria and West Nile virus were removed from the planet along with the hated mosquitoes, although a lack of citrus ushered in the return of scurvy. As hunger killed humans slowly, other maladies progressed.

Insects formed the basis of alternative medicine in various parts of the world, including India, Brazil, China, and swaths of Africa. Honey was used as an antioxidant and antimicrobial substance, deployed in the treatment of heart disease. Wasp venom was found to kill cancer cells. With the rise of antibiotic resistance, insects were once seen by researchers as a crucial source of new, widespread medicines. Perhaps they would even help beat back the next pandemic—after all, the Novavax COVID-19 vaccine was developed in altered cells of the fall armyworm moth. The catastrophe snuffed out such hopes.

Before long, the struts holding aloft most life on Earth were yanked away. Nearly 90 percent of wild flowering plants relied on pollination to prosper. Shorn of this service, and lacking the nutrients that insects recycle back into the soil, the plants died. Gardens became lumpen deserts. Wild meadows vanished, followed eventually by tropical rainforest trees. More than half the human diet globally came directly from those formerly flowering plants, multiplying the starvation rates. Entire ecosystems collapsed, accelerating climate change. Cascades of extinctions rippled through our denuded planet. For those of us left, the misery was finally complete.

1

An Intricate Dance

The question of how long human civilization would withstand the loss of insects is both hideous and unfathomable. Hideous because the collapse of arable farming and ecosystems could wipe us out within just a few squalid months, the biologist E. O. Wilson has predicted. Most of the fishes, mammals, birds, and amphibians would plunge into oblivion before us, followed by flowering plants. Fungi, after an initial explosion from the death and rot, would also die off. "Within a few decades the world would return to the state of a billion years ago, composed primarily of bacteria, algae and a few very simple multicellular plants," Wilson wrote.

And yet, unfathomable. Such a dire scenario can barely be comprehended given the stubborn survival of insects through the five mass extinctions that have roiled Earth in the past 400 million years. Humans have never existed without them, so have never had to properly consider their absence or even diminishment.

But a torrent of recent findings have pointed to major declines in the abundance and species diversity of insects in places around the world. Seemingly without cause they are crashing, their numbers thinning out at astonishing rates at different research sites—in some places by half, others by three-quarters, and in one, in the seemingly benign countryside of Denmark, as cataclysmic as 97 percent. The mounting evidence of plummeting insect populations forces us, for the first time in our history, to grasp the wretched consequences of their decline. This book will explore the unfolding crisis in the insect world, what's causing it, and what can be done to stem the loss of the miniature empires that hold life aloft on our raucous, plastic-strewn, beautiful planet.

In a bewilderingly rapid reimagining of our world, what was once infinite now seems jarringly vulnerable. Without insects, the world's wealthy could perhaps deploy the resources required to indefinitely stretch out a semblance of the status quo. But for the majority of humanity, the loss of insects would be an agonizing ordeal eclipsing any war and even rivaling the looming ravages of climate breakdown. "Most of life on Earth would disappear if we didn't have insects, and if there were any humans left they wouldn't be having much fun," says Dave Goulson, professor of biology at the University of Sussex. "I think it is stretching it a bit to suppose that all humans would be dead in a few months, but there is no doubt that millions of us would be starving."

Insects have been involved in an intricate dance with almost every aspect of the terrestrial environment for millions of years, forming an underappreciated foundation for human civilization itself. They multiply our food, act as food themselves for the other living creatures around us, rid us of the foulest waste, eliminate unwanted pests, and, crucially, nourish the soil, the 15-centimeter (6-inch) patina wrapped around our globe that sustains all of humanity. Rachel Warren, a professor of environmental biology at the University of East Anglia, compares our deeply woven reliance on insects to the internet. "In an ecosystem everything is connected by this net of interactions," she says. "Every time you lose a species you're cutting some of those links in this network. The more links you cut in the network the less of this internet there is left, until eventually it doesn't work anymore."

Without a pollinator, a plant dies and isn't replaced. The birds that feasted on the plant's fruits or the deer that browsed on its buds start to dwindle, followed by the animals that feed upon them. "The whole food web just disintegrates," Warren says. "I don't think humans could survive in that world at all."

The weight of this dependence has failed to spark much devotion for insects. Three out of every four known animal species on Earth are insects and yet, within their massed ranks, only butterflies are considered with anything close to affection. Wasps are a baleful summertime menace, ants an invading army fought with toxic sprays in the kitchen, and mosquitoes everything from irritating nuisance to lethal threat. Most of the other 1 million species of identified insects are considered by many people, if they are ever considered at all, to be either quirkily obscure or pointless.

There are around 7,530 types of assassin fly, a creature that spends its short life spearing other insects with a sturdy proboscis in order to paralyze them and liquefy their internal organs. This horde alone comprises more species than mustered by the entire world of mammals—apes, elephants, dogs, cats, domestic cattle, whales, the lot. A botfly called *Cephalopina titillator* matures in the nostrils of infested camels, just one specialist among 150 species of botflies, while there are at least half a million species of parasitoid wasp, creatures so detested by Charles Darwin that he wrote in a letter, "I cannot persuade myself that a beneficent and omnipotent God" would've created them. What would really be lost if these abhorred wasps and flies, maybe all flies in general, just vanished?

"You get rid of flies? You get rid of chocolate," says Erica McAlister, a senior curator at the Natural History Museum, London, and an avowed defender of flies who once took part in an entomologist go-kart event dressed as one. Appropriately, she successfully chased down a colleague who was dressed as feces. "Flies are really important pollinators when it comes to carrots, peppers, onions, mangoes and a lot of fruit trees. And chocolate. They work longer hours than bees and don't mind the cold as much. We're beginning to finally take notice of all this." There are approximately 160,000 species of Diptera—an order more commonly called true flies or two-winged flies—which includes houseflies, midges, mosquitoes, and fruit flies. The number of fly species is at least four times larger than all the different types of fish found in the world's oceans. This diverse group perhaps deserves to be viewed as a collection of finely tuned environmental engineers rather than as annoying pests that circle overhead or speckle browning bananas in fruit bowls.

Tiny midges, each the size of a pinhead, crawl into the tiny flowers of cacao plants across Africa and South America and keep the world's $100 billion chocolate industry from collapse. Thousands of different blowflies, flesh flies, and soldier flies dispose of dead animals, rotting leaves, and feces—for free. Scientists have harnessed maggots for the treatment of gangrenous wounds without antibiotics, while oil has been extracted from the larvae of black soldier flies and turned into a form of biodiesel to run cars and trucks. "They're doing such wonderful jobs, all sorts of things that we just don't realize," says McAlister. "Can you imagine if they didn't? You'd be swimming along in a quagmire of feces with Uncle Jeremy floating past you."

Flies are recondite yet prodigious pollinators. *Volucella zonaria*, a hefty hoverfly with bumblebee-like black and yellow hoops on its abdomen, is "basically a flying tank," according to McAlister. It is capable of buzz pollination, which means it can grip onto petals and violently vibrate, releasing pollen that is stubbornly lodged in the anthers of a plant. Few bees are able to do this, meaning without flies there would be no cornucopia of tomatoes and blueberries available for us to feast upon.

Some plants are completely dependent on certain flies. One extraordinary creature, *Moegistorhynchus longirostris*, is found on the west coast of South Africa. It has a nonretractable proboscis that measures up to 7 centimeters (almost 3 inches) long, several times its own body length, making for an awkward flailing appendage when flying. It flits around plants that have developed tubed flowers that perfectly fit the fly's lengthy probe, further highlighting an evolutionary theory posed by Darwin after he was sent some orchids from Madagascar in 1862 that stored nectar in exceptionally long necks. Darwin suggested that a moth with an absurdly long tongue must have evolved alongside this plant—a species that was only discovered decades after the evolutionary theorist's death. "If just that fly in South Africa disappeared, eight plant species would die out immediately," says McAlister. "Flies have got a huge history with pollination that has been wildly ignored."

Even on their own terms, flies can fascinate—some species present edible gifts to potential mates, while others perform intricate dances. To some people, flies could even be considered beautiful. Michelle Trautwein experienced a pivotal moment as an art student when as part of a studio review she unveiled a vast biological illustration of a stone fly, an order of insects with elongated bodies, long antennae and two pairs of membranous wings. "The art professor hated it," recalls Trautwein. The professor strongly preferred the work of a student who had smeared wet cat food across a blank white canvas. "I remember just thinking 'That's it. I'm out.'" Trautwein "just fell into flies" and is now a leading entomologist in her discipline at the California Academy of Sciences.

While stone flies are not typically gushed over as classically photogenic, there are flies that could lay claim to such adulation. The *Lecomyia notha* soldier fly, from Queensland, Australia, has an iridescent, opal-like exoskeleton, a shimmering blur of purple and blue. Another fly, with a bright, golden abdomen, has been named *Plinthina beyonceae*, after the singer Beyoncé. "Entomology is a really beautiful, aesthetically pleasing field," says Trautwein. She was drawn to flies, and insects in general, because they resemble "aliens on Earth."

"There's millions and millions and millions of them, we don't even know how many," Trautwein says. "Each one is like an alien life form with a detailed life history that often is so bizarre, you couldn't create it as fiction if you wanted to." As dizzyingly diverse as insects are, they share a remarkably consistent body design comprising three segments—head, thorax, and abdomen—three pairs of jointed legs, compound eyes, antenna, and an external skeleton.

This structure provides the platform for feats that would cause widespread awe if performed by larger animals. The dracula ant can snap its mandibles at 322 kilometers (200 miles) per hour, the fastest animal movement on Earth. Their cousins, the African Matabele ants, have been seen carrying injured comrades back to the nest to tend to their wounds like six-legged paramedics. Some caterpillars generate their own antifreeze to ward off the cold. Honeybees understand the concept of zero and can add and subtract numbers. But these creatures—so numerous that they are both unknowable and annoying, so odd looking that they inspire the forms of malevolent beings in horror movies, and so vital that we would perish without them—now appear to be suffering a silent existential crisis.

The alarm over insect declines has been rung intermittently for some time, if not quite as loudly as now. As early as 1936, Edith Patch, the first female president of the Entomological Society of America, gave a speech decrying the expanding use of insecticides on fruit and vegetable crops. "Certainly too little popular emphasis has been given to the service of insects to mankind," Patch said, adding that "too few do realize our dependence upon them for most of our food and clothing, a significant amount of our industry, and for much of our pleasure." More presciently, "If [mankind's] goal is a wholesale destruction of dangerous insects, his brains will provide the equipment for such a campaign in the course of time."

In the decades since, humanity hasn't consciously geared its collective brain to decimate all sorts of insects, much as it hasn't deliberately decided to drown its coastal cities and fuel enormous wildfires through climate change. Nevertheless, that has been the result. Through the destruction of insects' habitats, the spraying of toxic chemicals, and, increasingly, the heating up of the planet, we have unwittingly crafted a sort of hellscape for many insects, imperiling all we rely upon them for. "We are creating a world that is not only a problem for insects, but even for us, for humans," says Pedro Cardoso, a biologist at the Finnish Museum of Natural History.

The exact dimensions of the insect crisis have long been obscured by a fog of logistical impossibilities. There are 1 million named insect species, but as insects are small, cryptic, and not extensively tracked, this is only a glimpse of what is undiscovered and unnamed: estimates vary from an eye-watering 30 million species to a more realistic 5.5 million. "Who knows what's out there?" says Goulson. "Probably all sorts of weird and wonderful beasties."

Taxonomists, the biologists who name species and work out where they fit into the larger puzzle of living things, face a Sisyphean job just to differentiate between seemingly identical species. To most of us, some ants are black and some cinnamon colored, some flies are big and some are small, but beyond that the distinctions end. Specialists have to spend a lot of their time gazing at insects' reproductive organs to make their classifications. "We are genitalia fiddlers," says McAlister, the fly expert. "We like nothing more than cutting open a fly and looking at its goolies."

This painstaking work, combined with the fact that taxonomy is increasingly dismissed as a fusty natural history version of stamp collecting by students now more drawn to molecular biology, means that the job of describing all insect life on Earth will probably never have an end date. As McAlister puts it: "We've got 50,000 people studying one type of monkey and one person studying 50,000 types of flies." For every fly successfully identified by its genitalia, science dumps many more potential candidates on the desk of taxonomists. In 2016, Canadian scientists completed a DNA analysis of more than 1 million insect specimens and were shocked to find that the country probably has around 94,000 insect species, nearly double the previous estimate. If Canada has 1 percent of the world's insects, the researchers mused, the planet has around 10 million insect species.

Even with what is already described, it's clear we live in an invertebrate's world. Just 5 percent of all known animal species have a backbone. The globe is filled with not people or sheep or even rats, but beetles—350,000 species and counting. What we do know about overall insect populations doesn't immediately spur thoughts of shortage, either. The Smithsonian Institution estimates there are around 10 quintillion (that's a 10 with eighteen zeros following it) insects in the world.

A locust swarm can contain 1 billion individuals. The southern portion of England alone hosts 3.5 trillion migrating flying insects a year, a mass of bodies weighing the equivalent of 20,000 flying reindeer.

If you got all the termites in the world and scrunched them into a giant ball, this seething clump, a measure known as biomass, would weigh more than all the birds on the planet. Before people started ballooning in both population and girth in our era of industrialized modernity, all the world's ants probably weighed more than all the world's humans, too. "Today's human population is adrift in a sea of insects," as a pair of Iowa State University scientists wrote in 2009. "Based solely on numbers and biomass, insects are the most successful animals on Earth."

Insects are surprisingly hardy and adaptable, too. The Sahara desert ant can survive temperatures of up to 70°C (158°F), while, in the other extreme, the larvae of the Antarctic midge can cope with −15°C (5°F) and as long as a month without oxygen. Tiny ephydrid flies can live and breed in the hot springs of Yellowstone National Park that would fry a human. Bumblebees have been found at 5,500 meters (18,000 feet) above sea level, a height just shy of Mount Kilimanjaro's summit. Dragonflies can steadily hover in fierce winds that would down even the most advanced helicopter. A horned dung beetle is so strong that if it were a human, it would be able to hold aloft six double-decker buses.

You could say that the insect family embraces the bizarre. Insects breathe in and out via holes called spiracles in their exoskeletons and see via intricate compound eyes, allowing creatures such as dragonflies to have a 360-degree field of vision. Stingless bees feed on human sweat and tears, a species of butterfly has an eye on its penis, and some aphids can produce young that already contain their own babies—effectively they give birth to their own grandchildren. Insect populations are normally fairly elastic, too, able to navigate huge spikes and troughs when dealing with changeable conditions. But while insects are legion, that doesn't mean that they are utterly disposable—they all play some sort of role in pollination, or in decomposition, or in the food chain.

Start yanking enormous numbers of insects out from the environment and the whole web of life, including humanity, is thrown off-kilter. The collapse can fold in on itself, too—around 10 percent of insects are parasites, often of other insects. If certain wasps can't find caterpillars to act as their slave puppets and egg hosts, or if certain flies can't hijack an ant's brain and then decapitate it, they, too, are under threat. This dangerous scenario is now coming into focus as scientists have started to piece together the puzzle of insect life. A warning shot was fired in 2014 with a compendium of available research that found that a third of invertebrate species documented by the International Union for Conservation of Nature (IUCN) are in decline, with these population decreases amounting to 45 percent globally over the past four decades. The losses were nearly double that of vertebrates.

Almost all of Orthoptera, an order of insects that includes locusts, grasshoppers, and crickets, is on a downward trajectory, as is the majority of species making up the vast order Coleoptera, or beetles. "Such animal declines will cascade onto ecosystem functioning and human well-being," the study of IUCN data warned, framing this calamity within what's known as the planet's sixth mass extinction—the ongoing annihilation of nature, unprecedented since the demise of the dinosaurs, at the hands of the smokestacks and bulldozers of humankind.

This roiling extinction event has some formidable totems—tigers, rhinos, elephants, polar bears. The plight of these animals, often referred to by the unlovely term *charismatic megafauna*, dominates media discourse and conservation funding. The success or failure of the effort to halt the ransacking of Earth's biodiversity is regularly seen to hinge on the fate of the handful of large beasts that are endlessly portrayed in movies, advertising, stuffed toys, and sporting team logos.

This "institutional vertebratism," as the entomologist Simon Leather has put it, is, in a more literary realm, reminiscent of George Orwell's *Animal Farm*, where "all animals are equal, but some animals are more equal than others." We are drawn, moist eyed, to some species and withdraw with a shrug from others. Insects largely find themselves in the latter category.

Insects, along with the mollusks, worms, and other *sans* backbone creatures that comprise the vast majority of species on the planet, have been relatively overlooked by the world of science and the limelight of celebrity. Entomologists have attempted bursts of showmanship to turn the tide—a newly discovered species of treehopper was named after Lady Gaga due to the "wacky fashion sense" of its horns, a beetle has been named after Arnold Schwarzenegger, a wasp after Pink Floyd—but it is hard for many people to warm to insects.

Small children are fascinated by insects and want to interact with them, according to Scott Hoffman Black, executive director of the US conservation group Xerces Society, who does regular outreach in schools. But this attitude changes by the time they are in middle school. "Many actively fear, dislike or are disgusted by insects," he says. "I believe this is something that is taught by parents, peers and even teachers." The media's treatment of insects has played a role, too. In 2020, the United Kingdom's annual emergence of swarms of flying ants out to find a mate was greeted by the *Liverpool Echo* headline "Flying Ant Scenes 'Like a Horror Film' as Swarms of Insects Plague Merseyside." Children were reported to have screamed in terror, while one man compared the scenes to a Hitchcock film. We've been taught to fear the abundance of nature, when the reverse should be the case.

We didn't know what we were losing because we didn't really care or perhaps because we simply didn't know what was at stake. Neglect and ignorance became confusingly knotted some time ago.

Then, seemingly from nowhere, everything changed. The public's awakening to the insect crisis has come in waves and is far from complete, but it can plausibly be traced back to an exact date—October 18, 2017.

On that day, *PLOS One*, an open-access scientific journal headquartered in San Francisco, published a paper authored by a dozen Dutch, British, and German scientists. Its title was workmanlike and to the point: "More Than 75 Percent Decline over 27 Years in Total Flying Insect Biomass in Protected Areas." The paper itself gave flesh to these bleak bones. A rare long-term study of insect populations in sixty-three protected nature areas across Germany revealed a cataclysm: since 1989, the annual average weight of flying insects caught in traps slumped by 76 percent. The situation at the height of summer, when insect numbers reach their apex, was even worse, with an 82 percent decline.

Changes in weather and land use couldn't account for the overall drop, according to the paper. Despite being in protected and often actively managed conservation zones, it appears that the insects were harmed by activities on surrounding farmland, such as the use of pesticides and loss of flowering borders, although this theory of a sort of "ecological trap" isn't conclusive. A more pressing question, however, was if insects are seemingly plunging into oblivion in protected areas in a country like Germany, where on Earth could they possibly be safe?

The tone of the researchers was dark. Hans de Kroon, a Dutch ecologist involved in the study, remarked: "We can barely imagine what would happen if this downward trend continues unabated." Fellow researcher Goulson gave it a go, regardless: "We appear to be making vast tracts of land inhospitable to most forms of life," he said, adding that future generations are set to inherit a "profoundly impoverished world."

The findings ricocheted around the globe, prompting not just unprecedented interest in the struggles of flies, moths, bees, and butterflies, but also a wave of biblical language. "Warning of 'Ecological Armageddon' after Dramatic Plunge in Insect Numbers," read *The Guardian*'s headline. *The Hindu* went with "Insect Apocalypse: German Bug Watchers Sound Alarm." The *New York Times* also invoked "insect Armageddon" before, for good measure, declaring in a magazine piece a year later "The Insect Apocalypse Is Here." A cover of *National Geographic*, crawling with pictures of beetles and moths, declared mournfully, "You'll miss them when they're gone."

The public was introduced to the portmanteau "insectageddon," which quickly took hold across the media. The reaction swelled to a despairing crescendo; in an article for *Le Monde* entitled "Compassion for the Weevil!" the philosopher Thierry Hoquet intoned that "by chemically attacking insects, it is life that is attacked."

Much of this attention was heaped onto the unflashy membership of the Krefeld Entomological Society—composed largely of working scientists in various fields (the widely used epithet "amateurs" tends to irk)—that collected the data for the study that was structured by a group of Dutch, German, and British scientists. By the time yet another camera crew showed up, this time from the Australian Broadcasting Corporation, the society's insect curator, Martin Sorg, told them that all the commotion had been "problematic." Sorg admitted, "We never expected to get so many emails and so many questions from around the world."

Sorg, with his long gray hair, John Lennon glasses, and penchant for rumpled clothing and sandals, has become the unwitting face of both the study and the emergent concern over plunging insect numbers. He is a circumspect man, however, who is mildly bemused over why no one else has really bothered to make long-term standardized surveys of insects until now. "It's like we are driving a car blind," he says. "You may be lucky doing that or maybe you aren't. The less information there is, the more risk. I don't know why we were the only ones."

Since the era of intrepid Victorian insect collectors, scientists have strived to answer compelling questions about insect behavior or uncover intriguing new species. The drudgery of actually trying to count such boundless numbers—going to and from traps, compiling the figures, somehow supporting this work for decades beyond three-year research funding cycles—felt both pointless and dull. "There are so many interesting things to do that it sounds quite boring to do that," says Vojtech Novotny, a Czech ecologist who spends half his year conducting research among huge stick insects and butterflies in the Papua New Guinea rainforests.

Suddenly, however, Sorg and his colleagues are much like the only people who saw the need to keep score in a football game that everyone else only belatedly realized was important. The life's work of these assorted insect obsessives has centered on an old school building in Krefeld, a city in northwest Germany once famous for producing silk. A ribbon of the Rhine River cuts through the landscape a few miles to the east, and the Dutch border is not far to the west. The Krefeld Society has been trapping, observing, and recording insects since 1905, its members churning out several thousand publications during this time on the taxonomy and behavior of the animals.

The second floor of the group's building is given over to specimens suspended in alcohol-filled bottles—Sorg estimates there could be 100 million insects, maybe more—labeled and stored in disused classrooms, heavy curtains blocking out the light. In a separate part of the collection around a million insects have been dried, lanced on needles, and placed in frames. Butterflies, beetles, bees, hoverflies, dragonflies, and more, from the Rhine region and beyond, are here.

Crucially, the researchers have erected identical traps in the same controlled conditions across the countryside for the past four decades to ensure a clean comparison. The contraptions are called Malaise traps, named after the Swedish entomologist René Malaise, who developed the basic design in the 1930s, and resemble hovering tents open at two sides. The structures funnel flying insects up to a well-lit point where they are trapped in alcohol, forming a pile of bodies weighing a few grams (about half a teaspoon) each day.

Year after year, the Krefeld team collected and noted the mass of insects in the same nature reserves, which are meadows full of birds, small mammals, and wildflowers but set in a patchwork of agricultural land across Germany. Then, in 2011 and again 2012, they noticed something was amiss. "There was a place that should have a high number of insects, over 1,000 grams [35 ounces], and it had 300 or 350 grams [12 ounces] for the complete year," Sorg says. "That was shocking." The society's records on insect abundance spans technological ages, running from handwritten notes to documents bashed out on typewriters to saved files on floppy disks. By digging through these records, Sorg and his colleagues could see that numbers were well down compared with 1989, one of the earliest years for the standardized traps.

So they set about, with the help of the outside scientists, piecing together the deteriorating fortunes of the insects. Previously regarded as a group of niche eccentrics, they gathered evidence of the most significant disappearances of creatures since woolly mammoths were cleared from the continent 10,000 years ago, perhaps even since the demise of the dinosaurs. Yet the huge declines documented in Germany weren't completely revelatory to Sorg. He and other entomologists had muttered to each other about dropping numbers for some time. Even in the dustier tomes stored in the former school, society members had noted decreases before the Second World War. "We just did not suspect it in this dimension," Sorg admits.

The latent insect crisis was now visible as another sorrowful example of environmental pillage. "Until the Germany study most of the general public were completely unaware that there was any problem at all with insects, and also largely unaware that insects had any value," says Goulson, who started studying bumblebees in earnest in the 1990s after noticing, to his horror, that many once-common species had evaporated from southern England. "It's nice to see that it's not just a few sad entomologists that are concerned now. People are starting to wake up."

The Krefeld work is notable in that it measures biomass—a handy way to track changes in the bulk of insect life and speedier than the exacting work of identifying and counting each caught bug. But the method also provokes further questions. If the overall weight of trapped insects fell, is this due to a drop in larger individuals, such as bumblebees and the heftier beetles, while everything else is relatively steady? Or are all species nose-diving? Are whole species being lost or just portions of them?

Sorg argues that the focus should be on the "irreversible loss of species" rather than merely biomass, pointing to how the Krefeld region used to have around two dozen bumblebee species, as documented a century ago. This list has since halved.

Extinctions are a cruel blow to our sense of well-being with the environment. They remove irreplaceable threads from the tapestry of life, depriving us of creatures that perform important functions or make the world a more lively, interesting place. Lost insects such as the Perrin's cave beetle or the Xerces blue butterfly may not garner the fame of other departures, such as the dodo, but they were unique and their exit is irreversible.

The hidden, labyrinthine nature of arthropods—a broad phylum that includes insects, spiders, and centipedes—makes it alarmingly easy to stamp out whole species. Move across a patch of ground and you'll see seemingly mundane features—a pile of dead leaves, a rock, a tree—but in reality, those are the micro-sized homes of a riot of insect species. Glance from the ground upward, from the soil to the barks of trees to the canopy of a forest, and you are taking in yet additional strata of insect habitats with countless more species.

If this plot is flattened to construct a Starbucks or an intensively farmed field of soybeans, plenty of common insects perish—and so do niche habitat specialists. Some of these rarer insects may exist elsewhere; others may not and are removed from our world. The scale of unseen insect life is so broad and deep that it's difficult to keep track of the extinctions, let alone population fluctuations, unleashed as we blunder obliviously around the planet like some sort of intoxicated moose in a field of precious orchids.

Countless insect species have undoubtedly been extinguished without us even knowing they existed in the first place. These Centinelan extinctions, named after a ridge in Ecuador at the foothills of the Andes where a cornucopia of new species were wiped out before they could be named, have left researchers groping in the dark at the full scale of the insect crisis.

We may well already be within the first or second act of insect extinctions. A paper by twenty-five researchers ominously titled "Scientists' Warning to Humanity on Insect Extinctions" notes that only around a fifth of the world's insect species have even been named, mostly from single specimens. But by using a formula based on the extinctions of land snails and on previous work by Claire Régnier, of the French Natural History Museum in Paris, the report states that

5 to 10 percent of insect species have gone extinct since the era of mass industrialization. This range equates to 250,000 to 500,000 lost insect species, meaning that the tiny burst of geologic time since the arrival of the steam engine and incandescent light bulb has been an era of doom for up to half the number of species that have existed during this time and have been named by science. "We are pushing many ecosystems beyond recovery, resulting in insect extinctions," the paper states. "Insect declines lead to loss of essential, irreplaceable services to humanity. Action to save insect species is urgent, for both ecosystems and human survival."

We are perhaps better equipped to ascertain potential future losses than past extinctions, although this is of cold comfort. A landmark United Nations finding in 2019 outlined how 1 million species across the animal kingdom are facing extinction in the coming decades. Half of these lost species will be insects. In aggregate, this means that a period bookended by the late nineteenth century and the midpoint of the twenty-first could see the permanent disappearance of a million different kinds of beetles, butterflies, bees, and other insects. This toll, if it plays out, is gargantuan—a loss of species greater than all the variants of fish, birds, and mammals in existence.

But the loss in overall insect numbers matters as well, perhaps as much as the number of species being extirpated. As the warning paper from twenty-five scientists notes, declines are not restricted to rare and endangered species. The ranks of common insects are being thinned out, too, which has profound consequences for the surrounding environment.

Pull different levers and you set in motion a cascade of consequences. Across the broad family of arthropods there are creatures such as wood lice, millipedes, and springtails that perform tasks such as chewing up dead plant matter, grazing on fungi off root surfaces, and releasing nutrients for plant growth. Waste-eating insects such as dung beetles unlock nutrients from feces, rotting plants, and corpses that would otherwise stagnate. Other species like ladybugs and lacewings prey on crop pests such as aphids. The engineering aptitude of termites—their tunneling cracks open hard ground, helping it absorb water and nutrients—can help turn barren land into fertile fields.

If whole species of these specialists are lost, then vital ecosystem functions, such as maintaining soil and plant health, are diminished. But certain animals themselves feed on these insects in huge volumes; a blue tit parent, for example, will need to cram up to a hundred caterpillars a day down the gullet of a single chick. The loss of a few niche species won't bother most birds if they are able to dine on other insects that boast strong populations. A major crash in overall insect numbers, though, is a different matter. We can marvel at the individual qualities of insects, but their role in the ecosystem is almost always executed in massed numbers. It's not just about the breadth of the insect universe, it's the depth, too.

Insects aren't being persecuted in isolation, of course. The UN report identifying 1 million at-risk species also found that three-quarters of the planet's land has been radically altered by human activity, that plastic pollution has increased tenfold since 1980, and that the globe has been shorn of a third of its forested areas in the industrialized era. Our presence now hangs so heavy that we are starting to notice that it weighs us down, too. "The essential, interconnected web of life on Earth is getting smaller and increasingly frayed," said Josef Settele, who cochaired the UN assessment. "This loss is a direct result of human activity and constitutes a direct threat to human well-being in all regions of the world."

This shriveling of biodiversity is an emergency now on a par with, or arguably greater than, the climate crisis that feeds and overlaps it. The recent rush of academic warnings on insects has more than one parallel with how climate change has developed as an issue—a few largely ignored alarm bells eventually followed by a belated critical mass of concern once we near the precipice of disaster. We may now be edging toward this climax of unease. Biologist Pedro Cardoso has long obsessed over spiders and insects—he's a particular fan of parasitoid wasps: "Their way of living, often controlling the mind of their hosts, is way too cool"—but his past decade of studying insect declines has often been rather lonely. "It can be a bit frustrating when all the attention goes to mammals and birds," he says. "It's really the small stuff that drives what's happening in ecosystems."

Lately, though, Cardoso has noted a change. He will be peering through some vegetation in Ghana or sweeping a bug net around Finland and local people will come up to him to chat about the things they don't see anymore. The conversations are mini bouts of unprompted mourning for the ladybugs that used to gather here, the butterflies that used to flutter there. "This is coming from unexpected people," Cardoso says. "People who didn't even realize that they even cared about insects." He adds that "this flood of papers that are coming out helped our cause because finally people are realizing what's happening." The scale of activism over insects still doesn't yet compare to climate change, of course. "Maybe we can get a kid like Greta Thunberg," Cardoso suggests.

Nearly a year after the bombshell Krefeld study was published, another piece of research surfaced into view that was described by one entomologist as "one of the most disturbing articles I have ever read." It appeared to show that the insect crisis reached beyond Europe to the Americas.

In the mid-1970s, Brad Lister, an ecologist now based in upstate New York, took a research expedition to the rainforest of Puerto Rico to document its insects as well as their predators—the birds, frogs, and lizards. The El Yunque rainforest, located on the slopes of the Sierra de Luquillo mountains near the eastern tip of the island, is a lush carnival of biodiversity, boasting the endangered Puerto Rican parrot, the trilling noises of coqui frogs, and a tangle of different snakes. Lister needed a waterproof jacket for the trip—El Yunque is deluged with an estimated 605 billion liters (160 billion gallons) in rainwater a year.

To catch insects, Lister would use rudimentary sticky traps not far removed from the days of Alfred Russel Wallace and Charles Darwin. The researcher would smear a pile of plastic plates with Tanglefoot, a sticky compound, and distribute them on the forest floor and into the canopy. At sunset the plates would be a blackened mass of insects, ready to be picked off by torchlight, before being dried and weighed. "It was a lengthy procedure back then," Lister recalls. When he returned to the rainforest thirty-five years later to follow up this work with his colleague Andrés García, an ecologist at the National Autonomous University of Mexico, he could immediately tell something had changed. Or, rather, vanished. Large puddles that once featured flocks of butterflies were now devoid of life. Few birds zoomed overhead.

When it came to replicating the sticky plate tests, the suspicions deepened. "After the first day Andrés said 'where are all the insects?' and I said 'good question' as there didn't seem to be anything around," Lister says. "There were signs that something was amiss."

While in the 1970s the sticky plates ended up matted in insects, this time they regularly came back with just a couple of sad specimens. This happened day after demoralizing day. Once the results were published, the appalling comparison with the first research trip was clear. On the ground, 98 percent of insects by biomass had gone. Up higher, in the leafy canopy, 80 percent had disappeared. "It was astonishing," Lister says.

The duo also trapped anole lizards, thin green reptiles with fiery red throats, and found that the aggregated mass of the caught animals had dropped by more than 30 percent since the 1970s. This suggests, Lister says, that the rainforest is in a state of collapse due to "upward trophic cascade." This is a bottom-up version of a typical trophic cascade, where the removal of a dominant predator, such as a wolf or tiger, causes the food chain beneath it to become warped and the surrounding environment to alter.

The loss of insects, in contrast, was like removing too many blocks from the base of an ecological Jenga tower, causing the pieces above it to topple. Birds, frogs, and lizards simply had nothing to eat and so their populations shrank. El Yunque has been a protected place since the days of Spanish colonial rule, and so Lister and García felt able to rule out human interference, such as chemical-laden farming, for triggering the decline. Instead, they point the finger of blame at the heating up of the planet.

When the study came out, Lister initially thought the coverage was tainted with hyperbole. "The *Washington Post* wrote 'insect apocalypse' and I thought 'oh come on now, my name is ruined,'" he says. "But now I think that may be on the money. I've gotten a lot more radical. We are looking at a global collapse of insects and we have yet to sense the urgency of what that means for us."

The third dispatch in a ghastly trifecta arrived just a few months after Lister and García's research came out. A published analysis by two Australia-based scientists—ecologists Francisco Sánchez-Bayo and Kris Wyckhuys—made the boldest claim yet of a catastrophe for insects across the world that has few parallels, even throughout the long sweep of life on Earth. The metastudy's startling central finding is that 40 percent of insect species are declining globally, with a third endangered and at looming risk of extinction "over the next few decades." The rate of extinction among insects is eight times faster than that of mammals and birds, the analysis states, with the total mass of the world's insects receding at a breakneck speed of 2.5 percent a year.

According to the researchers, who looked at seventy-three reports of insect declines around the world, the orders Lepidoptera (which includes butterflies and moths) and Hymenoptera (bees, wasps, and ants) have been the worst hit, along with dung beetles. Aquatic insect orders such as Odonata (dragonflies and damselflies) and Plecoptera (also known as stone flies) have "already lost a considerable proportion of species," the paper says.

The study is an international carousel of despair, noting the loss of bumblebees across the United States, declining butterfly numbers in Japan, disappearing dung beetles in Italy, and dragonflies stripped from streams in Finland. It also uses blunt, apocalyptic language rarely seen in a peer-reviewed scientific paper. "Unless we change our ways of producing food, insects as a whole will go down the path of extinction in a few decades," states the paper, which blames the "dreadful state" of insect biodiversity on the destruction of habitat, pesticide use, invasive species, and climate change. "The repercussions this will have for the planet's ecosystems are catastrophic to say the least."

Of greatest alarm, according to the paper, is that the insect crisis is sweeping aside not just specialist creatures that rely on restricted habitats or particular host plants, but also "generalist species that were once common in many countries," suggesting there are broad pressures bearing down on all insects rather than isolated hot spots of decline. The precipitous decline of insects outlined in the paper is placed within the context of the developing mass extinction of many other species, but the researchers say that this collapse outstrips contemporary comparison and even the extinction event that snuffed out the dinosaurs 66 million years ago.

The paper claims that it is "evident that we are witnessing the largest extinction event on Earth since the late Permian and Cretaceous periods," a statement invoking a period dating to 252 million years ago. This was perhaps the most dreadful time to be alive during Earth's history, with a series of volcanic eruptions believed to have caused a huge extinction event known as the "Great Dying." Up to 96 percent of marine species were wiped out, along with 70 percent of terrestrial vertebrates. It was also the worst, and perhaps only, mass extinction of insects to date. The potential repeat of such a cataclysm should "prompt decisive action to avert a catastrophic collapse of nature's ecosystems," the paper warns, citing insects' sheer numbers and myriad roles in keeping life thrumming along for almost everything else on the planet. "It is indeed a period of crisis for all insects, even though we still don't know much about the status of certain groups because no one has studied them," says Sánchez-Bayo, who already knew that some insect groups, such as bees and butterflies, were facing challenges but acknowledges that it was a "major surprise" to find that beetles, dragonflies, and other insects were also in such peril.

This sketched-out tale of disaster further propelled alarm in the media, and increasingly the public, that insects face possible obliteration. Entomologists, startled to find their field thrust into the mainstream news discourse, started poking around their own dusty piles of population data or simply felt belated validation for what they had privately discussed for years.

Sebastian Seibold was one of these scientists. Seibold was part of a team that spent the best part of a decade until 2017 looking at the health of biodiversity in nearly 300 grassland and forest sites in the German states of Brandenburg, Thuringia, and Baden-Württemberg. Then the Krefeld study came out to huge fanfare, and the team realized they were also sitting on a trove of insect data. "We were sitting there reading it and saying 'Well our time isn't that long but we have this nice sampling, let's take a look,'" says Seibold, a researcher at the Technical University of Munich.

In the forest sites, the researchers had used flight interception traps, which consisted of a transparent plastic barrier stretched between trees that flying insects crashed into, sending them through a funnel into a collecting jar underneath. In the grasslands, sweep nets were used in June and August of each year. The work was time-consuming—each collected insect had to be taken to a laboratory in Munich, stored in ethanol, and then separated into its group. More than 1 million arthropods, spanning spiders as well as insects, were collected and sent to taxonomists. The team was able to group the specimens into 2,700 different species.

The trends in how the insects fared over the decade were surprising, Seibold says. His colleague Wolfgang Weisser had another word for it: "frightening." Across the grasslands, the number of species was cut by a third, while the overall biomass of insects nose-dived by two-thirds. In the forests, species numbers also declined by around a third, with biomass slumping 41 percent. Perhaps unsurprisingly, the grasslands surrounded by arable farming fared worst in the study, but in terms of species, there were drops across the board—carnivores, herbivores, and decomposers. In the forests, everything but the herbivores declined, but this was mainly due to a broader change in the mixture of trees, where conifers made way for broadleaves.

This time, the media was primed for the disastrous findings. The study came out at 7 p.m., nearly two years to the day since the Krefeld thunderbolt, and within an hour it was on national TV news in Germany. Over the next few days, it was picked up by major newspapers in Germany, France, Switzerland, and Austria. As he was deluged by tweets, it became apparent to Seibold that people felt something dear to them was in peril. "Tigers and rhinos are of course beautiful animals, but they live somewhere else," he says. "What people can do is care about what's living in their garden, or in their region. It's important that people are aware of what they do with their everyday decisions."

The barrage of attention-grabbing insect research caused some critical muttering in the research world about unscientific levels of hysteria. Declaring an emergency when less than 1 percent of insects have been assessed by the IUCN for their conservation status—compared with two-thirds of vertebrate animals—can seem premature. Seibold's research, like the other studies, contains caveats and unknowns. It doesn't tell us what the rate of insect decline is outside these German states, and ten years isn't a huge period of time anyway—what if the beetles and flies and bees all rebound in the following decade?

But let's ponder this work in tandem with the Krefeld study. In one swath of Germany the mass of insects has shrunk by three-quarters while in another, albeit over a shorter time span, the biomass drop is almost as bad, and one in three species has vanished. In any other walk of life—in medical testing, aviation safety, school test results—such horrendous trends would trigger a chain of emergency interventions. With insects we have waited, as if for Godot, for the next study.

Regardless, the insect crisis, in one form or another, was now out in the wilds of public discourse. The plight of the things that scuttle at our feet and hover in our gardens could never be overlooked again. Panning across the world, it became clear that plenty of evidence was just waiting for someone to point at it and shout loudly.

2

Winners and Losers

Overturn a stone in the garden and there's a chance you'll find a few ants, perhaps even a wood louse. Poke around in the bark of a tree and you may surface a spider or beetle. Scientists, in their own, more measured way, are also trying to unearth what is happening in the world of insects. What they've found so far has often been distressing.

In the United States, the abundance of four species of bumblebee has plummeted by as much as 96 percent in recent decades, with the bees' geographic ranges shrinking by nearly 80 percent, according to an analysis of thousands of museum and field samples. In 2017, the rusty patched bumblebee, assailed by the conversion of prairies and grasslands to farmland, urban sprawl, and roads, became the first bumblebee to be officially listed as endangered by the US government. This wasn't for a lack of contenders—the Franklin's bumblebee, for instance, is only found in a narrow strip of southern Oregon and northern California and hasn't been seen since 2006.

Bees that rely on specialized habitat seem to be struggling. The blue calamintha bee, its abdomen a striking metallic blue, was thought to have been completely wiped out from its home in a sand ridge in central Florida before being rediscovered recently. The ridge provided one of the oldest remaining scrub environments in the state but has been almost entirely obliterated for agriculture and housing. The bee is now squeezed into a home universe spanning just 41 square kilometers (16 square miles).

Head over the border to Canada, and the population of the American bumblebee, *Bombus pensylvanicus*, is now 89 percent smaller than if you had visited a century ago. Lots of other insects are receding from the country, too, with an official at the Canadian National Collection of Insects admitting that there are "thousands of species that have just disappeared from the collection—things that haven't turned up for years."

A short hop back south to the United States, in a protected forest in New Hampshire, scientists discovered that beetle abundance has "dropped steeply" since the mid-1970s, with an incredible average decline of 83 percent. Nineteen beetle families vanished completely. The number of different kinds of insect families, representative of species diversity, was down nearly 40 percent. This rugged New England setting, in the White Mountains range, is one of the most pristine stretches of woodland in the northeastern United States, thick with birch, maple, and spruce. A riot of moths, wasps, and beetles are among the insects that flit around an environment also frequented by larger beasts, such as deer, bears, and moose. The beetle-seeking researchers set up nine window traps—essentially a wooden frame with a pane elevated around half a meter (2 feet) above ground with a trough of water with soap or antifreeze at the base. Beetles living on the forest floor are able to take abbreviated flights, a bit like a chicken does, and end up in the mixture after bashing into the pane.

This work uncovered incredible downturns. Whereas in the 1970s the traps routinely caught thousands of beetles from the Pselaphinae beetle subfamily, in 2016 the beetles had "completely gone," according to lead study author Jennifer Harris, now of Penn State but at Wellesley College during the research. "The decline was really colossal." The losses were worst in the lower elevations—a good 2°C

(3.6°F) warmer than the higher altitudes of the forest—suggesting, as in Puerto Rico, that in an area far from agricultural or urban interference, climate change is to blame.

Beetles perform a range of crucial roles in this and other forests. When a tree is felled, they help chew up and break apart the wood, allowing fungi to enter these spaces to aid decomposition. This allows the tree's nitrogen and phosphorus to be distributed to replenish the wider forest. Some beetles also prey on other insects, keeping their numbers in check. In this delicate dance of interactions, beetles devour collembolans, otherwise known as springtails, which help decompose leaf litter on the forest floor. Without beetles, springtails proliferate and decomposition accelerates to the point where the carbon storage of the forest floor is diminished. Springtails also feed on microbes that break down carbon. These relationships are complex, and there is still much we don't know, but the loss of beetles could potentially have ramifications for tackling the climate crisis.

"Beetles have innumerable roles in the forest, I can't think of another group that does what they do," says Nicholas Rodenhouse, a veteran biologist who worked with Harris on the research. The consequences of removing the vast majority of these insects from an ecosystem are slow to reveal themselves, but the obvious risks are a "radically disrupted food web," says Rodenhouse, who wistfully remembers as a boy finding luna moths in woodland and stag beetles in his back garden. "We are now living in a massively defaunted world, which is sad because it's less interesting, less colorful," he says.

This new world is also "functionally different" and degraded, Rodenhouse notes, although scientists are still working out the ramifications of this. The declines themselves had been apparent to US researchers decades before the Krefeld study, however. "When the German research came out a lot of people thought 'shit, I should've published my work,'" Rodenhouse says.

A string of other localized insectageddons reach across the continent. Ohio's butterfly population has dropped by a third over twenty years. Grasshopper numbers fell by a similar amount, in the same time frame, at a site in Kansas. In California, the monarch butterflies that migrate to the coast en masse each year now do so in numbers around 1 percent of the total recorded in the 1980s.

Seemingly invincible hordes are being felled. Mayflies, frail-looking aquatic insects with a pair of membranous wings, emerge every summer from their nymph state to form enormous swarms. These prodigious clusters can contain 80 billion individuals, thickening the air to the extent they are picked up on weather radar. These swarms are particularly large in the northern reaches of the Mississippi River and the Great Lakes, with some towns having to clear the roads of mayflies with snowplows. Thus, scientists who took a look through the radar data were aghast to see that mayfly populations had slumped by more than 50 percent since 2012 throughout the northern Mississippi and Lake Erie regions. The declines are likely a result of water pollution and "if current population trends continue, could cascade to widespread disappearance from some of North America's largest waterways," the mayfly study warns. Worldwide, the Sánchez-Bayo and Wyckhuys paper estimates that a third of aquatic insects—a group including caddis flies, dragonflies and water beetles—are threatened with extinction.

This is grim news for more than just this impressive category of animals, some of which have developed gills like fish. One type of water beetle, called *Regimbartia attenuata*, can even survive being eaten by a frog by swimming through the amphibian's stomach and crawling out of its bottom. Aquatic insects form a fundamental foundation to the food chain, consuming algae and dead leaves while nymphs and then, later in life, finding themselves on the menu for an array of fish and wading birds, dragonflies, and bats. They are also an important indicator of water quality, as pollution tends to drive them out of streams and rivers. Freshwater insects in the United Kingdom have actually increased their distribution over the past fifty years thanks to clean water regulations in the country. "Each species has a novel role in the environment, it never acts in isolation," says Corrie Moreau, an entomologist at Cornell University. "Think of every organism like a walking or flying rainforest—when we lose a species we lose all the diversity associated with it."

Insects in Europe are, from what we know, even more challenged than in North America. A review of 120,000 butterflies caught between 1890 and 1980, combined with more recent data drawn from millions of sightings, found that butterflies have declined by at least 84 percent in the Netherlands. The Dutch researchers glumly estimate that the true decline is probably even larger than this. A separate study of dozens of traps set in nature reserves in the north and south of the Netherlands found widespread losses over two decades ending in 2017. The annual rates of decline seem to indicate a steady march to oblivion—large moths down an average of 3.8 percent a year, beetles dropping 5 percent, caddis flies an eye-watering 9.2 percent.

Other groups—such as true bugs, an order including aphids and cicadas that has had its name hijacked in popular discourse for insects at large, and mayflies—appeared stable, but the conclusions are downbeat. With the biomass of large moths slumping 61 percent and that of ground beetles falling 42 percent, the researchers noted that the "results broadly echo recent reported trends in insect biomass in Germany and elsewhere."

It is Britain, though, where the world's most detailed records on insects are kept. The country's zealous interest in insects stretches back to the early 1700s, embodied by a group called the Aurelians, composed of poets and artists who marveled at the apparently miraculous transformations of larvae into adult insects. By the Victorian era, bug collecting was a highly popular and social pursuit, with hordes of enthusiasts descending upon the countryside armed with nets, some of them storing their catches under their stovepipe hats.

The image of an eccentric vicar with a passion for butterflies endures when we think of this era, although insects gripped influential imaginations well into the twentieth century. Butterflies feature in the work of novelist Virginia Woolf, moths in the poetry of Siegfried Sassoon. Winston Churchill and Neville Chamberlain, both British prime ministers during the Second World War, collected butterflies. Walter Rothschild, scion of the banking family, had a collection of fleas dressed in tiny costumes, including a tiny bride and groom.

When the craze for trapping and pinning insects gave way to simply observing them, the research effort took off in Britain in a way that didn't happen elsewhere. A combination of experienced entomologists and an army of energized volunteer "citizen scientists" have embroidered much of what we know about insect trends.

At the forefront of this effort has been Rothamsted Research, the oldest agricultural research institution in the world; it is situated on the grounds of a sixteenth-century manor in Harpenden, a commuter town north of London. It is famed for the Broadbalk experiment, an initiative to measure the impact of fertilizer on crops that has run since 1843—a world record for a scientific experiment.

The Rothamsted insect survey has run two insect trap networks continuously since 1964, originally focused on migrating insects such as moths and aphids, but now much more broadly based across many insect groups. There are around eighty light traps in the light trap network in any one year in the United Kingdom and Ireland, mostly run by volunteers but coordinated at Rothamsted. The light that these traps emit has a broad wavelength, and this is particularly attractive to passing moths. A wide array of other insects end up there, too—around 1,500 different insect species in all since the start of the surveys. Even more visually striking is a separate network of suction traps. These contraptions, sixteen in total dotted across England and Scotland, are like large upside-down vacuum cleaners 12 meters (13 yards) in height. A fan housed in the trap sucks air down so that any passing insects—primarily aphids—are pulled into a container.

The results of these endeavors have been unhappily illuminating. The total abundance of trapped moths declined by more than a quarter from 1968 to 2007, with losses worst in southern Britain at 40 percent. A more recent analysis of 224 million trapped insects by Rothamsted researchers found that over forty-seven years, moths have declined by nearly a third, albeit with periods of peak and troughs since the

1960s. Aphids dropped slightly, although the researchers considered the long-term trend as being relatively steady.

Counterintuitively, moths have plunged in UK coastal, urban, and woodland habitats, but not in agricultural areas. Increasing temperatures, driven by climate change, should have boosted overall numbers, too, since they introduce newcomers. The Jersey tiger moth, for instance, finds London warm enough these days to relocate there from its home in the Channel Islands. But even with the efforts of dedicated researchers and volunteers, there still isn't adequate funding to help fully piece together this sometimes confounding puzzle. Still, the losses themselves are plain to see. "We are losing species. Clearly that is tragic," says James Bell, head of the Rothamsted insect survey. "I think scientists would agree widely that insects are declining. There's no question about it. Absolutely no question."

Moths are often maligned as powdery vandals that enjoy chomping their way through the clothes in our wardrobes, which is a slanderous generalization—it's the moth larvae, not adults, that feed on clothes, and even then only a tiny portion of moths do this. The United States, for instance, has around 15,000 moth species, and just two of these will ever pose a threat to a woolen sweater or cashmere scarf.

While they are vastly overshadowed by our fondness for bees, moths are in fact crucial generalist pollinators that help sustain plants that bees overlook. Researchers found that nearly half of moths tracked in the English county of Norfolk transported pollen from dozens of different plant species, including several rarely visited by bees, hoverflies, and butterflies. Sadly, both the unheralded labor of moths and the more extravagant beauty of their cousins, the butterflies, face growing threats.

Rothamsted's James Bell used to enjoy watching fritillary butterflies, a family that derives its name from the Latin word *fritillus*, meaning "chessboard," but now they are vanishingly rare. He still sees white butterflies in his garden but not many commas, which have brown-flecked wings that conceal them as they hibernate in dead leaves. He feels as if the world of his childhood has radically transformed. "There was a time when you used to go on your bike and you swallowed insects—that just doesn't happen anymore," Bell says. "I used to be stung by a lot of wasps and I'm not stung at all anymore." Many people, when they orientate their minds to it, can come up with similar anecdotes that point to a vanishing insect empire.

Scientists like Bell are in a select group that can actually elucidate these suspicions by making sense of the dead and missing insects; yet most insect researchers struggle to get funding for their work, passed over time and again for yet another treatise on a large mammal. "This has happened for decades, and is why we still know so little about most insects on Earth," says Manu Saunders, an ecologist at the University of New England, in Australia. "It's a vicious cycle—to win funding, you're expected to justify the need for funding with evidence; but if you can't get funding, you can't provide evidence."

The paucity of such long-term research has led to disagreements over the scale of insect decline as well as several open questions over the consequences of losing chunks of the insect world. What happens to other larger species? What happens to the forests, streams, and even cities? What happens to food production? E. O. Wilson and others can make an educated guess of the level of catastrophe, but firm answers are yet to materialize. "If two thirds of British insect species went extinct, what would actually happen as a result?" ponders Bell. "I couldn't possibly tell you. I can only give you a qualitative answer that it's bad."

In the United Kingdom, evidence of the "bad" is piling up. Another study on moths, in 2019 by scientists at the University of York, found that the creatures are dropping in abundance by 10 percent each decade in Britain. There have been periods of dramatic boom and bust, with a 1976 heat wave causing a huge surge in moth numbers, only for consistent gradual declines to set in from the 1980s onward. Yet another moth study, from 2014, found that since the 1970s, 260 species had declined significantly, whereas 160 had increased significantly.

British butterfly numbers have nearly halved in the past fifty years, while more than twenty species of bees and flower-visiting wasps have completely vanished from the United Kingdom since the Victorian era. Other species are retreating into ever-shrinking enclaves; for example, the great yellow bumblebee, once found across the country, now clings on only in the far north and west of Scotland. A broader loss of insects that act as pollinators appears to be underway across Britain. Of 353 wild bee and hoverfly species, a third now occupy smaller ranges than they did in 1980, research has found, with these losses concentrated in rarer species. Insects that pollinate crops are vital to our food security, yet "substantial concern exists over their current and future conservation status," the research paper warned.

The average distribution of insects in Britain has fallen by 10 percent since 1970, according to a broad assessment by the United Kingdom's National Biodiversity Network, which noted that the country's "invertebrates and plants are clearly receiving less specific attention than mammals and birds" despite there being "growing evidence that insects are showing rates of decline that may be greater than other taxonomic groups."

Goulson, the University of Sussex biologist, labeled this neglect an "unnoticed apocalypse" in a 2019 report for the Wildlife Trusts that states that globally the abundance of insects may have fallen by 50 percent, or even more, over the past fifty years. "The causes of insect declines are much debated, but almost certainly include habitat loss, chronic exposure to mixtures of pesticides, and climate change," Goulson wrote. "The consequences are clear; if insect declines are not halted, terrestrial and freshwater ecosystems will collapse, with profound consequences for human wellbeing." This collapse isn't just tearing apart complex interactions involving other animal species, plants, and organic matter. It is also acting as a sort of weight that is flattening the insect world into a more homogeneous lump, where a riot of eclectic, fascinating species is being replaced with a smaller, and arguably blander, group of creatures that are equipped to survive the torments of the Anthropocene.

Scientists that map genetic diversity have found that the variety of genetic material among insects takes a hit, more than most other groups of animals, when there is a greater density of humans. Global bee diversity started plummeting in the 1990s, according to one study, with around half as many bee species now showing up in collecting efforts for museums and other institutions compared with the 1950s, when surveys counted around 1,900 species a year.

Even seemingly modest human interference in the planet's most remote locations is squashing insect populations. Researchers recently discovered that the introduction of European weedy plants to the far-flung islands of the Southern Ocean, near Antarctica, has made the insect life there more uniform. "We are homogenizing the environment," the entomologist Simon Leather says. "If you grow a lot of soy and you use herbicides, you are sending out a big signal that says 'come here soya specialists' which are pests, beetles and aphids that feed on soya. A diverse set of natural enemies tend to need more diverse habitats."

To recast the natural world in this way leads us not to a global extinction of all insects but rather the exit of those unable to cope with the changes we have wrought, including many insects hugely beneficial to human civilization. In their place we are witlessly creating favorable conditions for animals we tend to loathe. "We're not going to lose insects but we may end up with a planet full of cockroaches and mosquitoes," as Timothy Schowalter, an entomologist at Louisiana State University, puts it. "We may make the world untenable for ourselves but insects will survive."

It's useful, therefore, to think about the insect crisis less like a single downward sloping line on a graph and more like a lot of different lines, some holding steady, some zigzagging, and some even going upward, while many others, representing species we consider interesting or important, head southward. If the loss of certain bees and butterflies is counteracted by a boom in houseflies and locusts, this swap won't be widely welcomed even if overall insect numbers remain roughly the same. The numbers themselves only tell us so much. "This is the messy part of science that much of the media coverage ignores," Manu

Saunders says. “We think that the public want simple answers, but do they really? We don’t need to dumb science down to get people to pay attention.”

Even as science grapples with the consequences of these losses, the bleeding continues, mostly without intervention. This inertia is perhaps summed up best by a 2013 paper by David Lindenmayer, an Australian ecologist, that looks at instances where threatened species have been monitored for conservation purposes but suffered local or total extinction due to the lack of action to save them.

One of the most infamous examples of this is the Christmas Island pipistrelle, a small bat weighing around 3 grams (a fraction of an ounce) that roosted in tree hollows. The bat was once common on Christmas Island, an Australian territory in the Indian Ocean, but suffered an 80 percent population drop between 1994 and 2006. Wildlife officials who were monitoring the bats pleaded with the Australian government to establish a captive breeding program before it was too late, but instead, a committee was formed to consider options. Months passed. More monitoring took place. By the time permission was given to capture bats for breeding, just one bat could be located through its echolocation.

Researchers desperately attempted to catch the creature but failed. The bat’s last calls were recorded before falling silent on August 26, 2009. “It is quite possible that this is one of the few times that an extinction of species in the wild can be marked to the day,” notes the IUCN in its description of the species. The name of the Lindenmayer paper summing up this and other extinctions abetted by such dawdling is “Counting the Books while the Library Burns.” In an era of shredded biodiversity, the title resonates. Parts of the insect world are undoubtedly on fire, and, scarily, there are a lot of books left to count. “We should start doing something even if we don’t know everything,” Sebastian Seibold says. “If you wait another 10 or 20 years, it might be too late. I can’t imagine how a world without many insects would look like but I don’t want to see it.”

Within Australia, Lindenmayer has been particularly vocal about the need to maintain the mountain ash forests in the state of Victoria to prevent the extinction of the Leadbeater's possum, a rare, endemic marsupial that nests in the hollows of trees often targeted by loggers. This small possum is just one of a parade of Australian species marching toward extinction due to ham-fisted human interference: habitat has been rampantly cleared, invasive species such as feral cats tear through billions of native birds and mammals a year, and climate change is starting to sink its teeth into what is already the driest inhabited continent on Earth.

Until recently, though, insects weren't thought to be in any sort of danger in a country so festooned by flies that the action of swiping them away from your face is known as the Aussie salute. Indeed, Australia is home to one of the great insect conservation stories—the Lord Howe Island stick insect, a mighty beast as big as a human hand that's also known as a "tree lobster." The species was thought to have been wiped out by an invasion of black rats before a few were found clinging to a remote rock jutting off the east coast. They were bred back to viability decades after being written off as extinct.

But Australia's insect life now appears rather more precarious than previously thought. The Christmas beetle, a member of the scarab family so named for its festive shimmering red and green coloring, was once a common sight after emerging each November and December. In 1936, a local newspaper in Queensland reported that the insects had swarmed in such numbers that "the noise of their whirring wings in confined spaces between buildings was like the sound of a far-off aeroplane."

Many Australians grew up seeing fewer in number than this horde, perhaps two or three over the holiday period, but now parts of the country appear to be vacated by Christmas beetles completely. Anecdotal tales fret that the beetles have been largely lost in certain regions, although no thorough surveys have been conducted. There is stronger data, however, on the mountain pygmy possum, a cousin of the Leadbeater's possum, with scientists finding in 2018 that between 50 and 95 percent of the animals had lost their full litters of young, which had starved to death. Their main food source, the bogong moth, known for its long migrations to the alpine regions where the possums live, has suffered a population crash. Australia has around 250,000 species of insects but only a select few, such as the bogong moth, green carpenter bee, and Key's matchstick grasshopper—a creature that has a stance similar to the upward-facing dog yoga pose—are systematically monitored. But a new front now seems to be opening up against Australia's beleaguered wildlife, a threat that will reverberate throughout its unique selection of animals. "The worry is, if insect populations are in decline, so are the populations of larger animals such as birds and lizards who rely on them as food," says David Yeates, director of the Australian National Insect Collection.

Some of the most challenging insect hardships have been reported in Australia's tropical and subtropical northeast, an area of the country stuffed with a kaleidoscope of often monstrous insects. Here, in the rainforest that stretches in a band up the eastern coast, you can find the Hercules moth, the world's largest moth, which has a wingspan as wide as a dinner plate but no mouth—it lives off the food reserves gobbled up while a bulky caterpillar—and two false eyes in its rear to confuse would-be predators.

The wet tropics in Queensland is also home to the Cairns birdwing butterfly, Australia's largest with an 18-centimeter (7-inch) wingspan, and the stalk-eyed fly, its eyes popping cartoon-style from its head on elongated rods. Potter wasps fashion mud nests filled with paralyzed caterpillars as snacks for their emerging young, while fearsome green ants set up their own bases in the foliage of trees and shrubs, squeezing silk from larvae to bind leaves together and using similar levels of teamwork to pin down unfortunate victims for dismemberment.

This living museum of insects has provided a livelihood to Jack Hasenpusch, who started collecting insects and breeding the odd butterfly on his lowland rainforest property just north of the town of Innisfail around thirty years ago. Hasenpusch soon realized that his pleasant sylvan hobby could provide an income, too, and started the Australian Insect Farm, which he runs with his wife and son. The operation breeds a variety of insects for collectors—it is licensed to export several hundred a year—and deploys its collection for educational purposes. Visits to schools usually involves children clamoring to see the majestic metallic blue wings of a Ulysses butterfly, or the incredible half-meter-long (half a yard) *Ctenomorpha gargantua*, a colossal stick insect. "We thought it was the largest in the world but China's beat us now," says Hasenpusch, without bitterness. "It's the largest in Australia, anyway."

The giant burrowing cockroach, one of the more notable insects bred by Hasenpusch, is a surprisingly sought-after pet for Queenslanders. This sturdy creature, its brown plates of armor making it look a little like a walking helmet, is the world's heaviest cockroach at around 35 grams (a little over an ounce) and, true to its name, burrows a meter (about a yard) underground to build a permanent home where it will live out its ten-year existence. "They are more like a little armadillo, they are so huge," says Hasenpusch. "It's an impressive insect."

This rustic lifestyle, set in a prehistoric fragment of the wild, provided a boom in insects so consistently each year that Hasenpusch could feel its rhythms and be confident things would never change. The past five years or so, however, have been unlike any other. Hasenpusch will turn on his insect light, and where hundreds of bugs used to congregate, just half a dozen show up. Bee numbers are down, as are moths. The Christmas beetles are, Hasenpusch estimates, 90 percent down. "It's been shocking," he says.

The apparent nadir hit in 2018, when even the trees wouldn't develop seeds. This situation is a critical one for the entire ecosystem, which includes a local icon called the cassowary—a huge flightless bird that is the second heaviest in the world after the ostrich. The cassowary is primarily famous for a razor-sharp claw that is used as somewhat overhyped evidence that it is the most dangerous bird in the world to humans, but it is also a key disperser of seeds through the fruit it eats. An ecosystem crash would take down the cassowary, which in turn threatens the plants it helps propagate.

Hasenpusch is baffled by the declines. Although the broader region features farms that grow bananas, sugarcane, and papayas, none are anywhere near the insect farm. "It's pretty pristine bush land here, there's no real reason why the insects should be down," he says. "It'll be worrying if this keeps going on, I don't know what we will do. Find something else to do I suppose. My main worry is for the environment, though." Hasenpusch thinks a lack of recent rainfall locally is a factor, but is in contact with other hobbyist collectors and entomologists around Australia and is unsettled by what he's heard. "They've also noted that a lot of beetles are down, too, and nobody can explain why," he says. "I'm just hoping it is one of those cycles."

Prizing open the secrets of the world's tropics will be a pivotal moment in better understanding the scale of the insect crisis. The greatest diversity in insect species is found in tropical areas, although this manifold mass of life is largely unresearched, with countless species not yet named by science. Still, there is well-founded apprehension that climate change, habitat loss, and other degradations caused by industrial agribusiness are taking their toll.

There have been a few glimpses into this shrouded insect dominion. A study of nearly one hundred dung beetle species in the state of Pará, in the Brazilian Amazon, found significant declines following the 2015 El Niño climate event—where the eastern Pacific periodically heats up and influences weather patterns—with losses worst in fire-affected areas of the rainforest. A separate, longer-term study in the lowland rainforest of Costa Rica found a reduction in the density and diversity of caterpillars over a two-decade period. This decline, which researchers say was likely fueled by increases in extreme rainfall and rising temperatures, is also dragging down the natural enemies of the caterpillars, along with the ecosystem services they provide.

Daniel Janzen, an American ecologist, has been studying insect species in Mexico and Central America since the 1950s, regularly coming back to the same research site in Costa Rica since he first went to the country in 1963. Janzen and his wife and research partner Winifred Hallwachs, who has had a selection of moths and a wasp named after her, have meticulously documented thousands of species.

They also helped to establish Area de Conservación Guanacaste, a World Heritage site in the northwest region of the country. The loss of insects due to impoverished habitats and climbing temperatures is so obvious, Janzen says, that he has had chats with aging gas station attendants who could list the insects—crickets, mayflies, moths, midges—that have vanished since their teenage years.

"But I do not spend my life counting them, any more than you spend your life counting pedestrians or cars," he says. "But you notice when there are fewer." In a 2019 article for the journal *Biological Conservation* entitled "Where Might Be Many Tropical Insects?," Janzen and Hallwachs noted how they had seen insects disappear from the drying-out cloud forests perched on top of tropical mountains, as well as from the soils and waters of the lowland tropics. If we continue a "constant war with the arthropod world, along with the plants, fungi and nematodes, human society will lose very big time," they wrote, swatting aside the idea that further evidence is required to sound the alarm. "The house is burning. We do not need a thermometer. We need a fire hose."

But the difficulty in getting our arms fully around the scale of the insect crisis, or even if the situation should qualify as a crisis at all, nags away at some scientists. Not long after the alarm bells started to clang for the insects, a countervailing chorus started to urge caution.

Several scientists of varying hues lined up with two broad objections to the insect crisis narrative—first, that the research showing insect declines was either flawed or localized, and second, that the outcry over disappearing bugs verged on hyperbole and was therefore pernicious to good science. The dissenters started writing rebuttals and fired them off to scientific journals, some of which were the publishers of the insect decline papers in the first place. "The quality of some of these papers has been relatively weak, either due to misinterpretation of data or due to overzealous claims," a group of thirteen scientists wrote in an edition of *Insect Conservation and Diversity*. The public is now more aware of insect conservation, they wrote, but "this spotlight might be a double-edged sword if the veracity of alarmist insect decline statements do not stand up to close scrutiny."

The critique highlighted a few potential pitfalls in declaring an insect emergency. We know little about historical insect populations, so plotting long-term declines can rely on speculative baselines of insects' status prior to humanity's most zealous meddling. As it's virtually impossible to survey each individual insect, reliance is placed on samples that can be misleading. And as surveys have been geographically sporadic, who knows what's been happening with insects in uncharted places?

Another response took aim at media coverage, listing headlines such as the BBC's "Global Insect Decline May See 'Plague of Pests'" as pushing an "exaggerated and unlikely narrative." This criticism, penned by a trio of scientists in the journal *BioScience*, lamented that geographically restricted findings, primarily in North America and Europe, were being improperly extrapolated to show "doom and gloom" global declines that will do little to galvanize public support for insect conservation.

A third riposte concedes that "many insect taxa are unequivocally in decline across many regions of the planet" but cautions that most data come from human-dominated locations more unfavorable to insects anyway and questions whether the drop in numbers is any worse than for mammals, fish, and other creatures suffering from the broader crisis playing out among living things.

It's treacherous to declare universal insect suffering when surveys don't yet stretch around the world or while it's apparent that not all species are on the wane. Studies have shown increases among moths in the forests of Finland, pollinators in southeastern Spain, and the desert-dwelling ants of Australia. These strengthening numbers are "exceptions that prove the rule of general declines" according to Tyson Wepprich, a biologist at Oregon State University, but still they show that the tapestry of insect fortunes is a complex one.

Critics have poured particular scorn upon the study showing plunging insect numbers in the rainforest of Puerto Rico as well as the analysis finding 40 percent of insect species are declining around the world. Lister and García's Puerto Rico work puts the long-term collapse in insect abundance down to the stresses of climate change, but critics argue that two combined temperature records used in the research are suspect because they rely on a weather station that was damaged by Hurricane Hugo in 1989 and subsequently moved to a spot that gave warmer readings.

Timothy Schowalter, who has conducted research in the Luquillo rainforest for decades, says that the research also assumed his own surveys of canopy arthropods as representative of the wider forest, but this wasn't the case—he hadn't sampled trees randomly. Instead he targeted a particular tree species of the genus *Cecropia*, so the inference was faulty. Insect abundance in the rainforest is probably being subjected to a boom-and-bust cycle shaped by drought and hurricanes, Schowalter says. Perhaps surprisingly, hurricanes often boost insect populations by causing a rapid rebound in new vegetation. "I mean in some ways this study is a wake up call to get some data," says Schowalter. "It's just too bad that it's misrepresented our data to get there."

In response to this, Lister says that weather station data were corrected from September 1992 onward to make it comparable with previous readings, something that was factored into his research along with congruent temperatures taken from a separate, nearby gauge. He also claims that Schowalter's previous work implies that samples were taken randomly from trees.

"No one can deny tropical storms obviously have varying degrees of impact on forest ecology, but based on our research these are often ephemeral and superimposed on the relentless, ever present impact of climate warming," Lister says, adding that he and García "object as strongly as possible" to accusations they misrepresented the data. The opposing biologists do agree, at least, that further data are required from this and other regions.

The other contentious study, by Sánchez-Bayo and Wyckhuys, shocked many within the scientific community with its sharp language as well as its doomsday findings. A team of Finnish environmental scientists criticized the paper for lacking a measured tone, while also arguing that it had "lumped together" only studies that showed declines, therefore skewing the results. "If you search for declines, you will find declines," the Finns wrote in a response titled "Alarmist by Bad Design." Manu Saunders, another of the critics, says the analysis "should never have made it to publication."

A back-and-forth of letters attacking and defending the paper played out within *Biological Conservation*, the journal that originally published the research. The debate was mostly conducted politely, as is the wont of scientists, and stuffed with technical arguments about the misappropriation of data and statistics and sampling bias. But it was not a pleasant experience for Sánchez-Bayo, the target of much of this opprobrium. "There is no doubt that the revelation that insects are facing a crisis has provoked a nasty reaction on the part of those entomologists and ecologists that do not believe the change is real," he says, stressing that the "large majority" of entomologists agree with his work. Sánchez-Bayo insists that he and his colleague did not make any sweeping claims of consistent global insect decline; rather they simply reviewed the evidence out there. "Only those who cannot accept the facts may think we use hyperbole," he says.

This wasn't just an ego-laden scrap over professional rigor and scholarly reputations. Most, if not all, entomologists have noticed that insects are ebbing away and have struggled to drum up any recognition of this. "A key point that is often forgotten in this story: we already knew there was a crisis," says Saunders. "We have known for decades." But this doesn't mean there will now be a gleeful headlong rush toward terms like "insectageddon."

Reticence is something that courses through each branch of the scientific tree. Even in the realm of climate science, marked by three decades of increasingly dire findings, there remains some reluctance to agitate loudly over collapsing ice sheets or monster hurricanes. That isn't how scientists are wired. Chris Thomas, as president of the United Kingdom's Royal Entomological Society, says that the number of queries about insect declines from the public and the press are now greater than any other insect-related subject by a "large margin." But Thomas also put his name to a letter with two other scientists warning that "hyping-up" insect declines based on patchy or biased data "could ultimately backfire if it subsequently turns out that some of the claims have been exaggerated."

This stance is drawn from scientific rectitude, but the attritional and often poisonous battle over climate change has also left scars that only reinforces the desire for caution. "My concern is not whether insects are declining or not, because I think they, on average, are," says Thomas. "But if people say 'insects are declining by 70 percent' and then it turns out they're only declining by 20 percent, and everyone says, 'Oh. Well, that's all right, then,' I think it's similar with climate change—'Well, it might have been 5 degrees. Oh well, 2 degrees is fine.' "

We tend to compare different lived realities and potential outcomes, Thomas says. If we are primed to brace ourselves for a certain adverse impact, a lesser blow can feel like an acceptable fate. If 20 percent of insect species go extinct, this would be a disaster, but we would consider this a good result if we had been expecting that 40 percent or more of insects were on the brink of being rubbed out.

Saunders, who has contributed to several of the armageddon critiques, says she's received support from entomologists but also "aggressive criticism" from those who argue that talk of an apocalypse is useful because it obtains a public stage for insect conservation. She frets that the costs of gaining that spotlight, such as providing ammunition to conspiracy theorists or opening the way to intended misinformation, outweigh the benefit of raised public awareness. Already, the agrochemical industry has defended the current regimen of pesticide use on the basis that reports on insect declines are flawed. "This is the sort of damaging knock-on effect we should be most concerned about when we think it's okay to exaggerate science for public attention," Saunders says.

Flamboyant headlines in the media shouldn't surprise anyone, of course. "Let's face it, 'insectaggedon' is far more catchy than 'insects in Iceland decline,'" says Jasmine Janes, an evolutionary biologist at Vancouver Island University. "It shouldn't be. We—the public—should care about both. But it is, for many reasons." Janes worries that the act of stirring concern could backfire by causing the public, overwhelmed with dread over a problem that seems too huge to fix, to simply switch off. The alternative message is a modulated, if less rousing, one, Janes suggests: "There is evidence of some decline, we need to look into it further in order to plan our next steps."

It's possible there will never be a unified scientific consensus of an insect apocalypse, not least due to the overwhelming task of collecting data over decades on more than a million small, elusive species. The crisis among insects may become even more apparent, but it will retain its nuance. Not all insects are going to disappear: there will be winners and losers, and some conservation efforts will invariably pay off. If the worst does happen, perhaps we will be able to adapt in a sort of utopian technocratic dream where we somehow manage to partially quarantine our lifestyles from our surrounding environment.

This hesitancy to declare an insectoid catastrophe has chilling echoes of the ponderous response to climate change. While further research is perennially needed to comprehend the heating up of our planet, we disastrously failed to swiftly act on the information we already had. The basics of the greenhouse effect were understood by the Victorian era, with more recent decades deluged by comprehensive scientific warnings of growing urgency. Yet even now, with scientists able to measure to a precise degree exactly how much ice is crumbling away from Greenland or to produce maps of detailed clarity on how Bangladesh, southern Florida, and Shanghai will become partially marine environments, governments dither.

While the hideous fire and flooding of the climate crisis is starting to resemble a Hieronymus Bosch painting, the decline of insects is more like a partly hidden Picasso—invisible to us in parts, slightly misshapen and ambiguous in others. But the basic outline is there, and most expert viewers know what they are looking at. The question of when it's appropriate to raise the alarm publicly is a scientific one but also a deeply moral and pragmatic one.

Some of the experts are getting impatient with this debate. "Acting with imperfect knowledge is something that we all do all of the time, in our personal and professional lives," a trio of insect specialists wrote to the journal *Conservation Science and Practice*, citing the use of effective medical treatments for diseases that aren't completely understood. The letter points out that insect declines have been found on every continent bar Antarctica, adding that "although there has been some criticism of specific studies, the overall trend is clear and the broad geographic reach is perhaps the most dire feature of the current crisis."

What's required is a "swift response" that "need not wait for full resolution of the many physiological, behavioral, and demographic aspects of declining insect populations," the scientists state. Besides, they argue, pushing for measures to rescue insects is hardly calling for society to gulp down bitter medication. Connected habitats free of pesticides will boost a range of species as well as water quality and other ecosystem functions, while preventing the spread of invasive species will spare vulnerable crops. Action on climate change will benefit just about everything, everywhere.

If insects are protected and we end up with a more vibrant environment, coastlines largely where they currently are, and the retention of food abundance, how many people will care if some of the initial projections were a little strong?

Scott Hoffman Black, one of the letter's authors, never quite thought he would be in this situation. When he joined the Xerces Society, he imagined a career battling away on behalf of a handful of rare species, such as the Uncompahgre fritillary butterfly, which was on the edge of being snuffed out from its Colorado home, rather than face a full-scale war involving once-common insects. Growing up in Nebraska, Black owned a rumbling 1971 Mustang in what he calls "the last gasp of the muscle car." A youthful Black spent many hours cleaning the car of splattered insects, but when he went back home in the 2000s with his own children, he noticed that barely a single bug crashed into his vehicle. Being a scientist, he didn't draw much from this experience than the anecdotal but then started to see the research and was convinced. Black likens this current period to the treatment of climate science in the United States, where the response to an obvious threat has regularly been stifled. "There's good evidence that had we taken action in the 1980s we would not be here," Black says, in reference to the escalating damage caused by the climate crisis. "I think we're in the same place with biodiversity loss. We have to take action in this next decade."

The huge changes underway threaten to render Earth into a completely different state by the time his own children enter old age, Black contends. "Some of our previous issues that we've had to deal with are going to pale in comparison to the ecosystem collapse will likely happen if these trajectories continue to go down," he says. "We are seeing very steep declines in diversity, abundance and biomass in virtually every study that's out there."

The full picture of this insect crisis is gradually coming into focus. In a sort of successor to the Sánchez-Bayo and Wyckhuys study, a dozen scientists conducted the largest review yet of research on insect declines, sifting through 166 long-term surveys, not just those showing declines, from almost 1,700 sites. They found a smaller but still steep fall in terrestrial insect abundance, an average 9 percent drop per decade since 1990. Encouragingly, aquatic insects appear to have grown in number, by around 11 percent a decade, probably due to efforts to cut pollution levels in lakes, streams, and rivers.

Several entomologists put the rise in aquatic insects down to a recent rally from a very low base, but still, the analysis points to the nuance lurking behind any mention of an insect armageddon. In some quarters of the media, coverage of the paper even declared that the findings were encouraging, evidence that our expectations for insects, and the natural world more broadly, have hit a dismal nadir. After all, if the 9 percent decline each decade is accurate and continues unabated, once the infants of today reach their twilight years their grandchildren will be amazed at tales of seeing a bumblebee.

By November 2019, the swirling debate over the insect crisis is looming large as the world's premier insect experts gather in St Louis, Missouri. Entomology 2019, featuring 3,600 delegates drawn from more than sixty countries, is held under slate gray skies, a biting wind whipping in off the Mississippi River as attendees file into the America's Center, a cavernous, utilitarian 1970s convention space in downtown St Louis. The logo for the event is a jolly image of a fly zooming through the city's famous arch, and the local media coverage was one of casual mockery. "Bugs bug me," ventures a TV reporter from KMOV4 before he ventures inside to confess himself astonished to learn that a spider is, in fact, not an insect.

It's easy to see why Entomology 2019 could be portrayed as being a meeting of eccentrics. Delegates are mostly white males, with a widespread prevalence of beards, sensible footwear, and the sort of khaki-colored clothing that suggests that an emergency trip to a tree-lined stream to look for water boatmen could arise at any moment. An exhibition space features items for sale, including T-shirts reading "Keep calm, it's just a bug" and, inevitably, a mocked-up album cover depicting four beetles walking upright on two legs on the Abbey Road pedestrian crossing.

The conference's highlight comes when, in the wake of two years of fevered media reporting on the demise of insects, a gathering of leading entomologists have their say. The discussion, to a packed hall, is helmed by David Wagner, a University of Connecticut entomologist with a neat moustache and glasses perched on his head who looks like Woody Harrelson's more academic sibling. Wagner is one of the scientists who put his name to a missive urging caution over talk of an insect apocalypse, but here he seems genuinely concerned. Declines are cutting across different insect taxa, involving aerial, ground-dwelling, and aquatic insects, he says, from the Arctic to the tropics. "Oftentimes we are worried about rare species but this is different, we are looking at the decline of common species, things that are the nodes of food webs," he says. In a nod to the outcry caused by the Krefeld study—Martin Sorg sits cross-legged in the front row—

Wagner says "It wasn't until 2017 that we really got the attention of those outside our world. That was a wake-up call for sure."

Some speakers point to the long backstory of concern. May Berenbaum, a member of the entomological faculty at the University of Illinois for four decades, explains how she was part of a group that found pollinator decline in North America back in 2006, amid a rumble of worry over bees. Janzen then takes his turn to talk, Hallwachs by his side. Bearded and wearing a puffer jacket, he grumbles about the stage lighting before sharing a photo of his car on a bridge over a small creek in his native Minnesota when he was in high school in 1955. The front end of the car is almost completely covered in piles of insects, the beam of its headlights barely cutting through the mass. "That's the kind of insects I grew up with," he says. "That's the kind of insects that aren't here today."

The ecologist then launches into a tale of paradise lost in a corner of Costa Rica that contains as many insect species as the entire eastern half of the United States. He shares another photo, this time of a white sheet, lit on a moonless night in the Costa Rican rainforest in June 1986—it is plastered with an array of flying insects to the point that the sheet appears more brown than white. The next picture is of the same setup in the same conditions in May 2019. A few large moths and a handful of other small bugs cling to the sheet, which is otherwise bare.

To Janzen, the transformation is clear-cut and profound. The amount of cloud that typically circles the top of a forest-covered mountain has reduced considerably due to a rising heat that burns it away. A scorchingly hot day is beloved by the ecotourists who flock to Costa Rica but is a "death valley for organisms" on the mountain, he says. Army ants have marched up the slopes and have cleared away ground-dwelling creatures, with everything from birds to wasps suffering from the fallout.

For many of the experts, the bigger enigma is our indolence to save insects, rather than the accuracy of the declines themselves. "Insects are going down the drain rapidly," says Peter Raven, a botanist and ex-director of the Missouri Botanical Garden. "Insects can't get to the table, we need to defend them. So let's get busy." Wagner reminds the audience that they are the ambassadors for millions of species that don't get to vote or advocate for themselves. "We don't necessarily need to collect more data to take action," he says. "Rome is burning if you're an insect and there are a lot of low and no risk options we can do now."

There are signs that such alarm is spreading beyond the esoteric realm of entomology conferences, although in a world not short of competing disasters the task of harnessing public outrage is messy; people's lives are often hectic, attention spans are ever more stunted, and while empathy is a useful human quality to tap into, it often fails to stir action, especially if citizens don't consider the cause to be materially significant to their own lives. That many humans' starting position toward insects is one of distaste or even fear is a further hindrance. Our cupboards are full of an arsenal of chemicals designed to kill insects, while popular culture associates them with pestilence or otherworldly monsters. Even our language is geared against insects—we refer to them as "creepy crawlies," a rather insulting term, and accuse irritating people of "bugging" us. This isn't fertile public relations territory for entomologists.

There are still sparks of understanding, a sense of disquiet that windshields are clearer now, that outdoor lights aren't swarmed as often, that those skimmed headlines about bees probably don't bode well. It's harder to expect people to join up the dots between the spraying of some crops or taking a transatlantic flight with a situation where strawberries are hard to come by and the trees have fallen silent of birds, but maybe this general feeling of unease is enough to help put the brakes on the insect crisis.

Another, more likely, outcome is that we simply shift our expectations of the world as it serves up an altered version of itself. Middle-aged people now remember bug-smeared cars from childhood holiday trips, but this memory, and therefore the imprint of a "normal" setting of things, will age and die with them. Foods that are abundant and cheap now may become scarcer and more expensive, but after a period of dissent, coming generations will get used to blander, less colorful diets. The countryside may become moribund, quieter, and more pockmarked with waste, but we have coped and even thrived as postwar landscapes were flattened by industrialized agriculture, and so we will find a way to muddle through again.

Adaptability has been key to our imperial phase on this planet, but it also rewards a certain amnesia. Shifting baseline syndrome, a term that describes the gradual shift of accepted norms in the world around us, is perhaps most vividly captured in a 2008 paper by marine ecologist Loren McClenachan, who decided to delve into a trove of historical photographs showing fishing enthusiasts with their catches in Key West, Florida. It's a tradition in Key West, the last in the string of islands that protrude outward from Florida's southern tip, for customers on fishing tours to line up for a triumphant photo with their catch displayed on a "hanging board." McClenachan dug up photos from the 1950s that showed people posing with fish that dwarfed them in size, reaching 2 meters (6 feet 5 inches). Into the 1970s, the fish appear roughly the same size as the people who caught them. By the time McClenachan bought a ticket for a deep-sea day cruise in 2007, the fish on the hanging boards were comparatively tiny, less than a third of a meter (about a foot) in length.

McClenachan calculated that sharks caught in 2007 off Key West were, on average, around half the size they were in the 1950s, while the large groupers of fifty years previous had given way to much smaller snappers. The ongoing degradation of the coral reef system along the Keys has made it much harder for the area to support large-bodied fish. Yet the smiles of the people in the photographs are just as broad through the years, each era featuring grinning tourists delighted with their day's catch. The price of the boat tours, adjusted for inflation, hasn't changed much either, even though people are catching much smaller fish now. The expectation of fishing in the 1950s only had a certain shelf life—each new generation came with a new frame of reference, a new baseline. The natural world may seem to have shrunken to the old, but not to the young.

The same holds true for insects, which can even terrorize people in their trillions only for the experience to fade away into little-known history. The Rocky Mountain locust once moved in colossal numbers across the western half of the United States, bringing a sort of dystopia to rural communities in the late nineteenth century, the massed ranks of locusts blotting out the sun for hours. "They looked like a great, white glistening cloud, for their wings caught the sunshine on them and made them look like a cloud of white vapor," read one account. Once they descended, the locusts devoured every bush, tree, and cornstalk in view, stripping grass, leaves, and even the quilts thrown in desperation over vegetable gardens. The horde burst into farmers' houses, emptying cupboards and shredding the carpets. There were even reports of people having their clothes eaten off their backs.

In 1875, an immense concentration of locusts was calculated to cover an area larger than California. The species appeared invincible. But by 1902, the locusts had been tipped into extinction, possibly due to changes in farming practices or a lack of genetic diversity. Within recent history, a generation of Americans had been regularly assaulted by a biblical plague of insects, and yet people living in the western United States now would find this prospect unfathomable.

Entomologists now brood over thoughts of what may slip into the vacuum of the forgotten past, as well as the toil needed to reverse the tide. Encouragingly, insect numbers do fluctuate wildly in normal conditions and can quickly rebound due to their prodigious reproduction. A monarch butterfly can pump out several hundred eggs a day, a queen bee more than a thousand. Insects can recover; they just need the breathing space to do so.

An enduring rehabilitation will require us to do things that will have no concrete measure of success other than avoiding consequences that are objectively bad. Just to maintain what appears to be the status quo will take a sustained effort involving lots of large and incremental changes, many of them out of sight to most people, to the way we develop land, produce food, and generate energy. But before all that, we need to show on a fundamental level that we care.

3

"Zero Insect Days"

For Anders Pape Møller, it was always about the birds. Specifically, barn swallows. A particular thrill was the sight of a barn swallow on the move, a blur of motion as it deftly intercepted flying insects as midair snacks. It's a fascination that has led to Møller spending half a century studying the species.

The swallows, with their steely blue back and wings, cinnamon-colored foreheads, and deeply forked tails, add a dash of elan to the flat farmland of Denmark's north Jutland region where Møller, now 67 years old, grew up. Møller's long career in ecology began when, at age 15, he ventured from the family farm to ring the legs of birds with tags to track their life histories. The barn swallows were common and easy to catch and affix with bands, and Møller quickly branched out to study the habits of other birds, such as swifts and house martins, and sometimes poked around patches of vegetation for its denizens, noting a butterfly here, a ladybug there.

Life plodded along peacefully in Kraghede (the d is soft—"it's a bit difficult for English speakers, Spaniards can do it," Møller says). The region, neatly divided fields of wheat, rye, and potatoes dotted with whitewashed houses and the occasional stream, changed little as Møller headed south to embark on an academic career in Aarhus, Denmark's second largest city. But there was a nagging suspicion that something was badly awry.

Barn swallows have a prodigious appetite, with a breeding pair and their helpless offspring devouring around 1 million insects over the course of a season. As many as fifty or sixty pairs of swallows were found on each farm on the Jutland peninsula, suggesting an environment seething with vast numbers of insects. Møller remembers scores of carabid beetles fleeing hay bales as they were heaved into trucks by farmworkers, grass verges humming with bumblebees, his parents' farmhouse inundated by flies in the late afternoons.

Over the years, though, the insects seemed to be disappearing. Through the 1980s and 1990s, it didn't require biologists like Møller to notice the decline. "Most people in the countryside could tell there was a reduction in the number of insects," he says. Insect-eating birds like barn swallows appeared to be receding, too. "It was so obvious. It was easy to see without any measurements."

Measurements are the primary tool of scientists, however, so Møller set about thinking how to do easily repeatable surveys of insects and judge any related impacts on the birds. In 1996, he came up with an audacious yet simple idea for a scientific experiment: he would drive a car along the same roads again and again and count the number of insects that smashed into the windshield.

Møller picked some routes and set about traversing them in his creaking Ford Anglia, a model that had ceased production in the 1960s. With the help of a couple of PhD students, he leased cheap vehicles from his cousin's used car business—"these weren't Rolls Royces," Møller says—and drove them up to nine times a day down the same stretch of road before stopping to eagerly count the number of insects smeared across the windshields. He has done this from May to September every year from 1997 until now.

This means that Møller has spent over twenty years either trundling along a straight, featureless road or peering at the mushed guts of insects. The offbeat experiment has baffled local farmers, who tease

Møller that he is just cruising around on his summer holidays. "They don't consider my work to be work," he admits. "Even today many people shake their heads when I explain what I'm doing. People who work in biology are often considered a little bit odd."

To others, though, the experiment touched a nerve. Mentions of a crisis in the insect world, even if understood to a degree, can seem abstract or remote. But the memory of scraping dead insects off the car windshield stirs a realization that it's largely a task from a bygone age, something almost sepia tinted that dwells in childhood holidays from years ago rather than anything that motorists have to contend with now.

The lack of bugs on car windshields is becoming the accessible emblem of insect decline, much the way dejected polar bears are now a sort of shorthand for the climate crisis. Møller found himself listening to the same anecdote again and again. "A few different people told me about when they were growing up and went for summer holidays they had to stop the car several times so that the windscreen could be cleaned, to see through it," he says. "Now, that rarely happens." For his study, Møller chose two sections of road—one a 1.2-kilometer (three-quarters of a mile) strip in Kraghede and another longer stretch of 25 kilometers (15.5 miles) to the west in Pandrup. He cleans the windshield, fires the car up, and accelerates to 60 kilometers (37 miles) per hour, at which point he's in research conditions.

Møller has become keenly aware of when an insect collides with his windshield. Mostly it is small flies, such as mosquitoes and midges, but every now and then he'll hear the louder thud of a bumblebee or beetle. At the end of the study site, he pulls over, counts the marks on the windshield, and notes the weather conditions. He takes accompanying measurements using sticky traps and sweep nets—a scooped net attached to a 1-meter-long (yard-long) shaft—to check the abundance of insects in nearby fields. This work has been as grueling as it has been unorthodox, but the results, when published, were stunning.

Biologists most commonly document small, subtle changes that have occurred during their research. Møller found an earthquake.

In the more than twenty years of his car-based study, insect abundance dropped 80 percent on the smaller stretch of road. On the longer route it was a virtual wipeout—a 97 percent decline. Insects had almost been completely erased from a seemingly stable, unchanged swath of Denmark. Møller says that these compiled figures represent "dramatic declines," but they didn't unduly surprise him. When he started the study, he would regularly be picking the internal goo of up to thirty different insects off his windshield. More recently, the windshield would often remain utterly pristine throughout his metronomic journeys. "We had a lot of zero insect days," he says. "A lot."

Furthermore, as Møller's study noted, the "abundance of breeding pairs of three species of aerially insectivorous birds was positively correlated with the abundance of insects killed on windscreens at the same time in the same study area." In other words, as the insects vanished, so did the birds, probably due to a lack of food. The local ecosystem had been hollowed out, from its base upward. What's almost as remarkable is that the farmland Møller traversed is itself so unremarkable. It resembles much of the food-producing countryside in northern and central Europe, suggesting that these devastating declines are not unique to the northern extremities of Denmark.

In a growing list of countries, birds that dine mostly on insects, such as warblers, swallows, and bluebirds, are suffering deeper population drops than omnivores like crows and starlings. An analysis of bird trends across Europe found that insectivores declined by 13 percent between 1990 and 2015, while omnivores remained stable. Researchers say these declines are worst among farmland-dwelling species, suggesting that loss of grassland habitat and farming intensification is favoring more generalist birds. The diminishing of habitat and insects is robbing us of creatures we unquestionably cherish. "People might not give a damn about insects but they like pretty birds in their garden," says the University of Sussex's Dave Goulson.

A day after Martin Sorg and his colleagues released their game-changing study on the vanishing insects of the German countryside in 2017, a striking piece of research was published that appeared to dovetail with it. In little more than a decade, an estimated 12.7 million pairs of breeding birds disappeared from Germany, around 15 percent of the country's wild avian population. Athough the shrinking bird numbers affected rare species, mostly it had wrenched away common fixtures such as sparrows, goldcrests, finches, skylarks, and yellowhammers. The common thread joining them: insects. "Almost all affected bird species feed their young ones with insects," Lars Lachman, ornithologist at the German Nature and Biodiversity Conservation Union, told the German news outlet DW.

Across the border in France, frantic researchers have been in a state of near-mourning over the country's lost birds. In 2018, it was reported that bird populations across the French countryside had slumped by more than a third since the turn of the millennium. Everyday species such as the whitethroat, ortolan bunting, and Eurasian skylark have been decimated, with French biologists describing the situation as a "catastrophe." The use of pesticides on crops is a leading suspect, but researchers also pointed to the birds simply having no insects to eat.

Further north, in Sweden, researchers acoustically tracked the northern bat, which was once the country's most common bat species, and found that it is in full-scale retreat. Encounters with the bat dropped by 3 percent on average each year, corresponding to a reduction of more than half between 1988 and 2017. This "dramatic" decline was possibly down to a lack of moths, the bat's favorite food, the study authors wrote.

A lack of insect meals has been blamed for the decline of farmland birds in the United Kingdom, too. The spotted flycatcher, a specialist predator of flying insects, is now greatly diminished, and the red-backed shrike, which feasts on large beetles, has been extinct from Britain since the 1990s. Birds nesting in urban areas, meanwhile, are being suppressed by a lack of insect food, with researchers finding that urban great tits would need nearby insect populations to double to have the same breeding success as great tits in the countryside. "Insects are the cornerstone of healthy and complex ecosystems and it is clear that we need more in our cities," says Gábor Seress, lead author of this finding.

Clues of a mutual insect and bird death spiral aren't just restricted to Europe. In North America, the eastern whip-poor-will, a nightjar that is regularly heard chanting its name but rarely seen due to its stealthy camouflage, has been declining by more than 2 percent a year from its range in recent decades. Biologist Philina English was keen to find out why this was happening, so she compared the chemical signatures from feather and tissue samples of living whip-poor-wills with museum specimens dating back to 1880 to ascertain what the birds used to eat. The published results were clear—the modern population was "declining due to changes in abundance of higher trophic-level prey." Simply put, there are fewer large bugs around now compared with a century ago, leaving an impoverished food supply for whip-poor-wills and other insectivorous birds.

A possible paradox of the insect crisis is that it's not the insects that will ultimately bear the brunt of any coming catastrophe. They will continue, in an altered composition, while much of the rest of life on Earth flounders around, having suffered a foundational earthquake. Instead of using the framing of "insect conservation," we should perhaps be thinking instead about bird conservation or food supply conservation or even human conservation.

"No matter how roughly we treat the planet, we are going to vanish before the insects will," says Scott Hoffman Black, of the Xerces Society. "But what we will see is fewer or no birds in the sky. If you want birds, you need insects. If you want fruits and vegetables, you need insects. If you want healthy soils, you need insects. If you want diverse plant communities, you need insects."

The prime value of insects, at least in our own self-interested gaze, invariably centers around pollination. A leviathan global food production system has been polished and streamlined in every possible way by technology and yet still hinges upon bees, flies, and other tiny pollinators to ward off the specter of starvation. The darkest fears over the insect crisis rumble around our bellies—what will happen if the things that create our food die off?

Almost all flowering plants in the world are reliant to some degree upon pollinators, primarily insects but also other creatures, such as birds and bats that inadvertently shift pollen from the male to female part of the plant to help create seeds for the next generation. Dietary staples such as wheat, rice, and corn are pollinated via the wind, but most foods that provide a dash of natural color on our plates—avocados, blueberries, cherries, plums, raspberries, apples—require pollinators. In all, more than a third of food crops grown around the world need a steady stream of insect visitors to maintain them. Some countries, such as the United States, rely heavily on an army of managed honeybees to meet the levels of pollination demanded by modern agriculture's enormous volumes. But in most places, the reliable flow of fruits and vegetables rests on the slender shoulders of wild insects that are being crushed under the burly boot of human activity.

The wonders of mechanized farming and swift trade corridors have provided much of the planet with a rich, if unevenly distributed, bounty of food, but a loss of pollinators threatens to unravel this system in an era where the global population is set to grow to nearly 10 billion people within thirty years.

The laboriously named Intergovernmental Science-Policy Platform on Biodiversity and Ecosystem Services (or IPBES) produced the first global assessment of pollinators in 2016, drawing on more than 3,000 scientific papers. This endeavor threw up some weighty numbers. The value of food production directly affected by pollination reaches up to $577 billion each year, a haul that includes 1.6 million metric tons (1.8 million US tons) of honey produced by honeybees and $5.7 billion worth of cocoa beans, the main ingredient of chocolate. The report estimates there are around 20,000 species of wild bees that pollinate crops, assisted by an assortment of butterflies, moths, wasps, beetles, and various vertebrates.

While the report stopped short of declaring an imminent threat to this food supply, it did point to sharp losses among insects where the patchy data were strongest, with more than 40 percent of bee species threatened with extinction in some places in Europe. It also highlighted two unhappily contradicting trends: while the abundance of pollinators appears to be dropping, there has been a 300 percent increase in the volume of agricultural production dependent on animal pollination in the past fifty years. Agriculture is becoming more pollinator dependent, particularly in developing countries, at precisely the wrong time. "If we carry on with business as usual of a growing human population, growing meat consumption, clearing more and more land for agriculture, climate change; well, we are going to reach some sort of crunch point with pollinators," says Goulson. "There is going to come a point where crop yields start to fail and that is just inevitable unless we tackle the issue of pollination declines properly. I don't see any immediate signs of us doing that."

Just three years later, in the wake of the trio of high-profile studies on insect declines, a larger follow-up assessment by 145 scientists was noticeably more unnerving. The headline findings in IPBES's compendium of the overall living planet were stark enough—1 million animal and plant species at risk of extinction, four in ten amphibian species hurtling toward their doom, a third of coral reefs set to dissolve away to slime.

A data gap still remains when it comes to insects, the report explains in rather frustrated tones, but there is enough evidence to make a "tentative estimate of 10% being threatened." As insects make up around three-quarters of described living things, this would mean that nearly 14 percent of all animal and plant species face extinction—which equates to more than the million species the IPBES report settles on. The "decline in pollinator diversity is challenging the production of more than 75% of global food crop types," the report warns. This confronting scenario was emphasized by the UN's Food and Agriculture Organization (FAO), which used World Bee Day in 2019 to sketch out a previously unthinkable set of circumstances.

As bees and other pollinators retreat from landscapes around the world due to intensive farming practices, habitat loss, use of chemicals, and climate change, both crop yields and nutrition are set to degrade, the FAO warned. Without remedial action, vital vitamins and minerals, such as vitamins A and C, magnesium, zinc, and folic acid, will become harder for people to obtain as it becomes virtually impossible to grow fruits, nuts, and many vegetables at scale in some places.

This denuded food bowl will increasingly see pollinated fruits and vegetables substituted by staples such as rice, corn, and potatoes. The insect crisis is set to strip from many of us not only stimulating and interesting meals but also the right balance of nutrients, stoking the risk of various ailments. One study found that pollinator losses could cause so many preventable conditions, such as heart disease, that the world may face an extra 1.4 million deaths a year. It turns out that losing wild bees and other pollinators is disastrous for our health.

Farmers, particularly in poorer countries, will see their livelihoods diminish, and the broader environment will suffer, too. As crop yields fall and the mix of grown food changes, it's likely that more wilderness will be swallowed up by agriculture to compensate. Falling biodiversity will help cause yet further declines, a vicious loop that will make our world look a much drabber place. "There is a huge global pollination deficit, we are already facing this crisis," says Lucas Alejandro Garibaldi, an Argentinian pollination expert. "We will need more land, which means more deforestation."

Even the sight of a pretty wildflower, nearly 90 percent of which are pollinated by insects, will become scarcer. A lack of pollination will degrade the production of various important materials, too; cotton, for example, can be produced without the help of insects, but pollination vastly increases the weight of cotton bolls and therefore boosts crop volumes. The cotton industry, worth $25 billion a year in the United States alone, relies on bees, flies, and beetles as an economic imperative.

We will desperately attempt to prop up populations of honeybees, but the sweeping task of pollination often falls on other, little-known, creatures. In the United Kingdom, for instance, 4 billion migratory hoverflies move to and from the country each year, making them the second most important pollinator after bees. As well as helping propagate Britain's soft fruits, the hoverflies eat trillions of aphids, an agricultural pest. As pollinators continue to plummet, our food system will be reliant upon fewer and fewer species. We should be able to arrest the declines of a few key crop pollinators, according to Simon Potts, a bee expert at the University of Reading, but for the rest? "I think we've already overshot the point of return for a lot of them," Potts says. Our predicament, he argues, is a classic conservation tale of woe—we didn't quite realize what we were losing until it had slipped through our fingers. We could, he says, end up with a "tiny" group of effective pollinators for crops, perhaps just a dozen species. Parallel declines among noninsect pollinators such as birds and bats will worsen this growing deficit.

Richer nations will probably have the technological and financial muscle to weather this storm, but small-scale farming operations in Africa, Asia, and South America confront an era of extreme vulnerability, as do the surrounding communities that rely on them for their food. A cruel twist is that malnutrition hot spots around the world overlap significantly with those areas highly dependent on pollination for healthy foods. More than 2 billion people in developing countries are dependent on smallholder agriculture—this is where a loss of pollination will be felt most acutely.

The decline in pollination will manifest itself in several ways. An eclectic mix of insect species ensures a more reliable pollination of plants, but richer diversities also boost the quality and quantity of fruits and vegetables. Potts likes to share a series of comparative photos that shows a strawberry that has been insect pollinated, another that has developed from "passive" or self-pollination, and a third that has been wind pollinated. Only the insect-pollinated strawberry appears recognizable as a strawberry—the other two are so shrunken and misshapen they look like someone has gnawed away most of the plump fruit and tossed the remnants away.

The scenario is the same across different fruits. A study Potts was involved in analyzed different apple types in the United Kingdom and found that insect pollination improved output of the Gala variety by up to 2.6 million kilograms (5.7 million pounds) a year. Ongoing insect decline "could have serious financial implications for the apple industry," the research warned.

The escalation in hunger and malnutrition won't spring on us with the speed of a coronavirus-style shock, but it threatens to be deeper and more enduring. "Do I think we're going to get to 10 billion people on the planet? No, I don't. Things are going to decline too rapidly for that," says Douglas Tallamy, an entomologist at the University of Delaware. "It can't support what we have now. So how can it support three billion more?"

Another looming, but less obvious, ramification of insects' troubles is found within the medical world. Insects and their derivatives have been used by humans for their medicinal properties for millennia. When, in 322 BCE, Aristotle described honey as being "good as a salve for sore eyes and wounds," the practice of using insects as treatments had already been established in China for around 1,000 years. Around 300 insect species are used in traditional Chinese medicine, with beetle larvae deployed to treat liver cirrhosis and the Chinese black mountain ant ground up as a powder or tonic that purportedly aids the immune system.

In parts of India, cockroaches are boiled in a soup to treat urinary obstructions, while in the south of the country some people use mud from the chambers of anthills to apply to outbreaks of scabies. In Brazil, more than forty insect species are used in folk medicine in the northern state of Bahia, mostly administered by grinding the toasted bodies of insects and gulping them down in a tea. Ancient insect medicines were often useless, such as those linked to the "doctrine of signatures" theory promoted in seventeenth-century England, where the physical resemblance of herbs or insects to body parts was thought to be significant. Many people fruitlessly used hairy ants to reverse their baldness or stick insects to accelerate weight loss.

However, modern science has belatedly started to confirm that many of the traditional insect medicines show promise in treating a range of conditions. Digging into the properties of insect treatments, researchers have found bee venom to be a potential weapon against certain cancers or even dandruff, while honey, an antioxidant, may be useful in combating heart disease and skin problems. A mouthwash of propolis, the hive sealant made by bees, has promise as a remedy for high blood pressure and gum disease.

The natural world has been plundered for life-changing medicines for centuries, although the richest seam has always appeared to be in the realm of plants and fungi. After all, morphine came from the opium poppy, aspirin from the salicin found in willow bark, penicillin from the fungus *Penicillium chrysogenum*. Despite the rich history of insect medicines, also known as entomotherapy, "insect-derived products have yet to establish the recognition and market success of their plant derived counterparts," wrote Rutgers University researchers Lauren Seabrooks and Longqin Hu in a 2017 paper. This mismatch, the duo stressed, was partly due to negative cultural attitudes toward insects in the West, and scientists should not be deterred from utilizing this largely untapped resource. "Indeed," they concluded, "if given the proper attention, insect-derived substances hold great promise for the future of natural product drug discovery."

The long evolutionary history of insects and their enduring use in traditional medicine should provide a useful springboard for medical research into new insect-derived cures. It can potentially help address the nightmare of antibiotic resistance, a phenomenon described by the Centers for Disease Control and Prevention (CDC) as "one of the biggest public health challenges of our time." As germs develop the ability to overpower the battery of antibiotics designed to kill them, there are fears that "superbugs" could race through communities without effective treatments. This prospect is spurring researchers to turn to the potential held within the tiny exoskeletons of insects.

The immunity to pathogens that shields insects such as cockroaches and beetles could be key in devising new antibiotics. Already there have been successes. In 2018, researchers at the University of Zurich announced that thanatin, a natural antibiotic produced by the spined soldier bug, blocked off certain bacteria. This new finding will hopefully be marshaled into something that will be a "very welcome addition to the new medicines urgently needed for effective antibacterial therapy," said John Robinson, one of the researchers.

It's one thing to find that wasp venom can zap cancer cells or that the blood of blowflies has antiviral properties, but it's quite another to scale that up from a few insect specimens into widely available medicines. Scientists believe they can find a path forward on this; one option could be to insert the required genetic material into certain insects, such as crickets, that could be mass-produced.

But the insect crisis casts a long shadow over these potential therapies. The tragedy lies in wasted opportunities, of treatments being ripped away before we even knew we had them in our grasp. The unseen loss of insects, the sort of Centinelan extinctions that confound our understanding of exactly what we've banished from the living world, means we may have carelessly burned through revolutionary medicines via the mundanities of farming, urban development, and other staples of modern life. "If we lose species without finding out what those uses might be, we deprive ourselves of options that we might have," says Schowalter. "We'll never know."

Insects are the most accessible animals in our lives, beyond our cats and dogs, and they can sometimes feel a little too close when a tick burrows into the skin or a battalion of fire ants troop into the dining room. But they are also the most otherworldly and, arguably, most impressively tenacious beings we share this planet with. Ironically, many of the insects we purposely try to rid ourselves of are flourishing even as the species we value fade away.

Wasps, for instance, are the dastardly villains of picnics and the wielder of a painful sting, all without the endearing quality of making honey. While many people would happily finish off the last wasp with a rolled up newspaper, the impact would be profound. These insects are a pollinator of plants, with figs particularly dependent on some wasp species. They are also key allies of gardeners and farmers by preying on species deemed pests, such as caterpillars, aphids, and whiteflies.

Far from their image as brainless stinging machines, wasps have even demonstrated the sort of logical reasoning previously considered only in the realm of humans. A US study showed that the lowly paper wasp can grasp transitive inference, a logical arrangement whereby if A is greater than B, and B is greater than C, then A must be greater than C. Paper wasps can also recognize other individual wasps by looking at their faces. Our sorting of "good" and "bad" insects is therefore not only unfair but in many cases redundant. "If you're defining that purely in the western context of convenience and comfort, then good and bad isn't a very useful label," says Michael Siva-Jothy, an entomologist at the University of Sheffield. "Wasps are possibly the best example of a 'bad 'insect that, if it disappeared, would cause havoc."

Few alien invaders are less welcome in our homes than cockroaches, yet they, too, have supporters who point out that we would be at a surprising loss if they were to all be stamped out. Scientists have morbidly marveled at cockroaches' ability to become resistant to a cocktail of poisons, leaving them free to potentially spread an array of bacteria types, including *Escherichia coli* (*E. coli*) and *Salmonella*. Researchers at Purdue University, in Indiana, spent six months attempting to reduce the population of a group of cockroaches without success, leading the researchers to conclude in 2019 that the scuttling pests are getting "closer to invincibility." This looming nightmare can rather blot out the benefits these reviled insects bring to our existence. Cockroaches' poor reputation is largely down to two species—the American and the German cockroaches, creatures that have thrived in our sewage pipes, waste, and kitchens. These cockroaches get their names not from their country of origin but from Swedish zoologist Carl Linnaeus, who simply named the cockroaches after the places he got his specimens from.

The American cockroach is a particularly formidable nemesis, running at around fifty body lengths per second, which would be the equivalent of a human running at an extraordinary 338 kilometers (210 miles) per hour. Slow-motion video footage reveals that the cockroach can crash into walls at high speed with no loss of momentum before scaling vertically. These great survivors can fit into cracks as thin as a small coin, bite with a force fifty times their body weight, and survive for two weeks after being beheaded. They are coated in a waxy substance that prevents them from drying out in the heated and air-conditioned environments of our homes.

But the cockroaches that we curse as they scuttle across the kitchen floor are just a tiny fraction of the estimated 5,000 species of cockroach that science has identified, a multifarious kaleidoscope of life that spans the giant burrowing cockroach, which can measure up to 7.6 centimeters (3 inches) in length and is regularly kept as a pet in its native Australia, to the *Attaphila* cockroach, a tiny beast that dwells in the nest of leaf-cutting ants, occasionally hitching a ride on the back of flying queens like a miniature brown backpack.

The glow spot cockroach in Brazil has two lantern-like protuberances on its head that burn brightly at night, while cockroaches of the *Prospecta* genus mimic ladybugs to deter predators. There are hairy cockroaches, cockroaches that can flatten themselves enough to be imperceptible to the eye, cockroaches that are able to curl up into a ball. Cockroaches can be submerged underwater without fuss for nearly an hour, survive blasts of radiation, and go a month without food. This is a broad family of survivalists that deserves grudging respect rather than blunt antipathy.

"Numerically, there is no question that most of the cockroaches in the world are beneficial to us," says Coby Schal, an entomologist at North Carolina State University. Schal has spent the past forty years studying cockroaches and, after getting over some initial discomfort at reaching into their colonies, now spends his time marveling at the beauty of the colorful and ecologically useful members of this vast family. Amid the vast ranks of cockroach species, Schal estimates only ten could be considered bothersome to humans. One, the German cockroach, only dwells in places frequented by humans and carries plenty of potentially harmful bacteria, although there is little evidence that this has provoked illnesses in people. Still, allergens in cockroach feces can provoke asthma in children.

"If I could eliminate the German cockroach from the face of the Earth I'd say absolutely yes, because it has no function outside the parasitic association with humans," Schal says. "But cockroaches perform very important ecological services. I've come to admire them." Even the German cockroach holds some appeal to Schal, even though working with the species has triggered allergies in him. "It's the most despised cockroach but it has adapted so perfectly to live in the human environment," Schal says. "We throw all sorts of insecticides at this cockroach and it develops all sorts of mechanisms to defeat that. You have to admire an animal that can adapt so quickly to human interventions."

Scientists have recently sought to exploit this indefatigability for human advantage, with mounting success. Cockroaches produce certain proteins to ward off harmful microbes, a defensive mechanism that could hold the key to developing a raft of new medicines for people, such as drugs that help overcome antibiotic resistance. In 2010, researchers at the University of Nottingham, in the United Kingdom, set about the edifying task of grinding up cockroach and locust brains. They found that the brain tissue contained an antibiotic agent that was able to kill more than 90 percent of methicillin-resistant *Staphylococcus aureus* (MRSA) and pathogenic *E. coli* without harming human cells. Separate studies found that cockroach compounds are toxic to breast and liver cancer cells.

In China, cockroaches have been used in traditional therapies for thousands of years. As if in a sort of dystopian fever dream, there is a climate-controlled facility in Xichang, in southwestern Sichuan Province, that houses 6 billion cockroaches. There are 28,000 full-size cockroaches per square foot in this cockroach farm, with the sound of them scurrying around likened by visitors to that of the wind rustling a pile of leaves. Officials have admitted it would be "catastrophic" for the surrounding area if the legions of cockroaches managed to break free from this compound.

Cockroaches are farmed in China as ground-up ingredients for a sweet, tea-colored "healing potion" that is given on doctors' orders to patients suffering from respiratory and gastric problems. The Chinese government has spent several years funding research into the medicinal benefits of cockroaches, with recent studies showing that they can help regenerate damaged tissue such as skin and the surface of internal organs. This is good news for people with stomach problems and burns victims—if they can get over their squeamishness at the source of their medicine. For such voracious destroyers and consumers of the natural world, some humans can be rather picky over which animals should be ingested. "I think we can value insects without cuddling up with them," Michelle Trautwein says.

It is an indictment of our era of separation from nature that we are so marooned from an understanding of what insects do for us, as well as the ties that bind us to them. Animals such as hedgehogs, lizards, and frogs certainly appreciate cockroaches—as a nutritious snack. Once you start hacking away those foundational, unloved links in the food chain, problems cascade upward, ultimately engulfing us. We are inextricably looped into this web even if the scaffolding of modern life—with its food delivery apps, cheap supermarket chicken, and nature repackaged as ecotourism—provides the illusion that we somehow float above it.

We may casually wish mosquitoes, for example, would simply vanish, but without them, many animals that we ourselves depend on would be denied a primary food source. Mosquito larvae are feasted upon by fish ranging from guppies to goldfish, and once the mosquitoes reach adulthood, they then become part of the terrestrial ecosystem, preyed upon by bats, birds, turtles, and dragonflies.

Ridding the world of mosquitoes would, of course, lift the burden of unbearable disease across the world's tropics. Since 2000, around 2 million people a year have died as a result of mosquito-borne ailments such as malaria, dengue fever, and yellow fever. By contrast, snakes kill a mere 50,000 of us a year. Mosquitoes cause people more suffering than any other animal by a considerable margin. Several scientists have argued that the end of this misery justifies the elimination of all mosquitoes, even if there is an ecological impact. Fish and birds would find other food. The makers of bug spray could find another line of business. We could all safely flop into camp chairs at dusk after a day hiking in the woods and not be darted by female mosquitoes seeking bloody nourishment for their eggs.

But there is a case for the defense of mosquitoes. There are around 3,500 species of mosquito but only around 10 are responsible in any significant way for disease in humans. Erica McAlister, the fly expert at the Natural History Museum, points out that mosquitoes themselves have never technically killed anyone—she has calculated it would take 440,000 of them to somehow bite and drain you of blood at the same time for you to die—but that diseases have taken advantage of mosquitoes' need for a blood meal and used them as a vector. "Malaria and other mosquito transmitted diseases are very bad, but we shouldn't get rid of all species of mosquitoes," she says. The vast majority of mosquitoes perform an array of ecological functions but receive even less credit for this than the other flies and cockroaches. Supporting the termination of all mosquitoes is "like saying let's get rid of all the primates, all the orangutans, all the gorillas, because I know one species of primate that's quite bad," McAlister says.

Mosquitoes are surprisingly adroit at pollinating certain plants, including orchids and tansies, due to their desire for floral nectar, although this is rarely witnessed, as they do this at or after dusk. Some researchers have theorized that to mosquitoes, certain flowers smell similar to us. They may weigh less than a grain of rice, but the dead bodies of mosquitoes do mount up, providing important nutrients for plants as they decompose. Mosquitoes have mastered some bizarre skills beyond biting, too—it's been discovered they have learned how to steal honeydew by stroking the head of a nearby ant, causing the fellow insect to vomit up its tasty meal.

These sorts of qualities have helped stir the odd, albeit rare, pang of empathy for mosquitoes. Catherine Hill, a Purdue researcher who spent twenty years exploring different avenues to kill mosquitoes, has described a moment of "epiphany" after her academic counterparts started looking into the potential of eliminating mosquito species using genetic modifications. "It occurred to me that in the last hundred years all we've thought about is how to kill mosquitoes and much less about their ecological role," Hill says. "We've been blindly following along in that mindset. Eradicating a species is a radical concept and I was uncomfortable with it. Alarm bells were going off."

Hours spent gazing at mosquitoes through a microscope gave Hill a different appreciation of them as "remarkable and beautiful creatures." This process triggered a shift in her lab research to develop an insecticide that will halt mosquitoes' ability to pass on disease, but without killing them. Other scientists are working on genetic modifications that will stunt the spread of mosquitoes. In one such project, dubbed a "Jurassic Park experiment" by critics, local officials in Florida agreed to the release of 750 million modified mosquitoes that will, via reproduction, produce far fewer biting females able to reach maturity.

Hill argues that the overwhelming focus on wiping out mosquitoes has trampled on any implications raised by their removal. Animals may find other food sources, but what if they don't? Other wildlife turn over organic matter and pollinate flowers, but how severe would the shortfall be? Hill cautions over these unknowns. "You pull one little piece and start to unravel it and things happen," says the entomologist, who, in perhaps the only endearing discovery ever made about a mosquito, revealed that the insects "sing" to each other via the flapping of their wings when selecting a mate. She says, "We have a very poor understanding of unintended consequences. Mosquitoes are complex creatures, like everything else, and we don't understand them quite as well as we think we do."

Hill's evolving view of mosquitoes has proved controversial with some researchers and businesses. But her reevaluation chimes with others who have spent their time studying insects rather than simply squishing them. "I get a bit sick of people coming up to me and saying 'what's the point of mosquitoes?' or 'what's the point of cockroaches?,'" says Floyd Shockley, who oversees the entomology collection at the Smithsonian National Museum of Natural History in Washington, DC. Cockroaches have a valuable job recycling plant material in nature, Shockley says, and without mosquitoes you impact larger invertebrates. Without those larger invertebrates, fish have nothing to eat. "At a certain point, you're going to reach something that matters to people," he says. "We didn't come from another planet, we live on this one and this is the only one we have."

Sadly, we are unlikely to witness a rehabilitation, or even vague reappraisal, of the reputations of anonymous or disliked insects. Their utility is overshadowed by aesthetic and cultural values as much as their supposed menace or uselessness, a situation that has left us woefully unprepared for the crisis that is now gathering pace among the massed ranks of insects. It is a crisis wrapped inside another, broader one. One million species on our planet are now at risk of extinction, a biodiversity disaster occurring at a terrifying rate that places growing demands on funding for research and conservation work.

Dams, roads, and palm oil plantations are scything through the rainforest home of orangutans. Wild tiger populations have recently stabilized but are still down around 97 percent compared with the start of the previous century. In 2018, Sudan, the last male northern white rhino on Earth, keeled over and died at the age of 45, leaving just two females left to represent the entire species. We are watching the extinction of large, charismatic creatures in real time, a terrible debt of human progress that will seemingly never be paid in full.

Amid this carnage, fretting over the loss of fireflies or beetles or even butterflies can feel incidental, even quaint. And yet the tragedy of wiping out rhinos would not threaten the viability of global food production, and the hideous crime of allowing all orangutans to perish would not provoke widespread child malnutrition, trigger the demise of dozens of bird species, or cause the landscape to be covered in rotting cadavers. In terms of impact, the insect crisis drowns out any other alarm bells in the domain of animals.

Insects' huge numbers make them appear both inconspicuous yet omnipresent. Even in the midst of a global nuclear winter, there would still likely be some ants and cockroaches amid the charred bones of humanity. "I think we would go first," said McAlister. "Insects survived all previous mass extinctions, after all." It is overwhelmingly for our benefit, rather than just theirs, that we should fret about the state of insects.

Insects' unshakable endurance has allowed them to play a background role in crucial points in human history. A Roman army once suffered a defeat in modern-day Turkey after becoming delirious from consuming strategically placed bowls of a type of honey, called "mad honey," that is hallucinogenic. Later, the Magna Carta, the US Constitution, and the works of Johann Sebastian Bach were all written using ink from oak galls, the small balls that form when wasps lay their eggs on trees. American independence itself was won with the aid of mosquitoes. The British army was so ravaged by malaria during the American Revolution that its southern commander, Lord Cornwallis, sought to withdraw to the north to avoid "the fatal sickness which so nearly ruined the army." Instead, he was ordered to hold Yorktown, where his depleted forces were overcome by American and French troops, prompting the end of the war. This collapse prompted the historian Timothy Winegard to call *Anopheles quadrimaculatus*, or common marsh mosquito, the "founding mother" of the United States. Fast-forward a couple of centuries and you'll find that a fruit fly became the first living animal in space when, in 1947, it was propelled beyond the atmosphere in a US military rocket in an attempt to ascertain the potential impact of cosmic radiation on astronauts.

This supporting cast of insects is rarely garlanded, but insects have been shaping our own story more profoundly than most of us would credit. A full accounting of insects' importance should prompt more sobering ideas as to our own position in the planetary hierarchy. The loss of insects would unleash a version of environmental armageddon, whereas if all of humanity were to vanish, the absence would barely be noticed by nondomesticated animals. Even our head lice would find new homes, on the primates we have driven to the brink of extinction. As the famous Thomas Eisner quote puts it: "Insects won't inherit the Earth—they own it now."

The human rampage through the ecological bounty of the world has its own era coined for it—the Anthropocene. E. O. Wilson has said he prefers to call it the Eremocene, or the Age of Loneliness. Even insects, those hardiest of planetary companions, now appear to be dissolving from our surroundings. This should disturb us far more than it currently does. "People say to me 'there are a lot of bugs out this afternoon, I'm not going to go out,'" Wilson says. "This is the sort of thing people say about what are, in fact, the little things that run the world. They run the world and we need them."

4

The Peak of the Pesticide

Alex Lees plodded along the plateau and surveyed the scene. To his left, a cluster of trees in an otherwise mottled brown landscape. In the middle distance cars trundled along the A628 road that slices through the valley below. Lees whipped out his phone, took a panning video, and opened up his Twitter account. "A precious fragment of @peakdistrict temperate rainforest, an island in a sea of managed #moorland bereft of #biodiversity," Lees typed out, posting the tweet at 11.12 a.m. on October 24, 2018. The downbeat message may have been surprising to some, given the Peak District's status as one of the most treasured and frequented national parks in the United Kingdom, drawing more than 13 million visitors a year.

It will also be a revelation to many that there are rainforests in Britain. Not the tropical type that blankets the Amazon, thick with jaguars, poison dart frogs, and sloths, but the temperate version found in the world's fringes—coastal Alaska, Tasmania, southern Chile. The wedge of trees captured in Lees's video is known as Middle Black Clough, a prehistoric ravine that features a tumbling waterfall. It is, like the other patches of temperate rainforest in Exmoor, in southwest England, and a larger swath in west Scotland, a relic from when rainforests were common to Britain following the last ice age, around 10,000 years ago.

These places are "fairy-tale" forests, Lees says, riven by gurgling streams, gnarled ancient trees, and imposing boulders blanketed in startlingly green ferns and mosses. Large, leafy lichens called tree lungworts actually resemble the lungs of these forests, its lobes splayed outward. Saturated by rain and the tang of humidity despite being in a mild climate, these scraps of rainforest conjure a Tolkien-like world where you could imagine a hobbit living within the hillsides or an elf fashioning a bow and arrow underneath the outstretched limbs of an oak.

Temperate rainforests, like most complex, wildlife-abundant wildernesses in the United Kingdom, have been receding for thousands of years and are now at the point of full-scale retreat, initially threatened by land clearance and now from the assaults of grazing animals and invasive plants such as rhododendrons, an import from Spain. "Rhododendrons bring me out in a kind of psycho rage," says Erica McAlister, the fly expert. "I go around trying to kill all rhododendrons in the UK, much to my mother's anger."

The tiny pocket of forest in the Peak District is a remnant of a bygone age surrounded by denuded rolling hills of peatland and limestone, now mainly used as pasture for sheep and cows or for the cultivation of silage for animal feed. This jewel of British nature is, to most insect and bird life, a wasteland devoid of food or shelter. "If I wanted to take someone to see wildlife in the UK I wouldn't take them to the Peak District because it's terrible," says Lees, an ecologist at Manchester Metropolitan University. "You just don't see wildlife."

The national park, partially sandwiched between the cities of Manchester and Sheffield, isn't absent of interest or beauty—walking trails offer sweeping views of greenery, stately homes here were used as filming locations for *Pride and Prejudice* and *Jane Eyre*, and an impressive cave network was inhabited by people, including the occasional bandit on the run, into the twentieth century. But as an ecological entity, the Peak District has been greatly impoverished in a process of ruthlessly intensive land management that is now well entrenched across the United Kingdom, Europe, and North America. Trees and scrubland have been torn down, bogs drained, wild meadows flattened. This has been a disaster for the insects packed into the barks of trees, buried in leaf litter, or crawling around the stems of tall grasses.

"We simply don't see large beetles in the countryside anymore," says Lees. Butterflies have also dwindled, and birds like the spotted flycatcher, which eats flying insects, are "going down the pan right now," Lees says. "That's happened in my lifetime and it's really worrying."

The Peak District felt like a pristine wilderness to Lees during childhood visits, but he realized later that this was because he was comparing it with the featureless arable land of Lincolnshire, where he grew up. Human meddling over thousands of years was staring at him in the face; it just took a moment to see it. "In the UK habitats have changed for so long and so much, no one has an idea of what biodiversity should look like," he says.

Lees lives about 100 meters (109 yards) from the boundary of the national park and leaves his windows open on warm days. "Virtually nothing comes in, no crane flies, nothing," he says. This sort of phenomenon isn't unique to the Peak District. "I'm turning 40 and I remember as a child moths all around the car," Lees says. "Now you might get some clouds of midges in the western highlands of Scotland but that's about it for real insect bioabundance experiences in Britain."

Britain may command the reputation of a land of enchanted forests, but axes have been ringing loudly here since the Bronze Age and have barely stopped. Half of the country's ancient forest has been vanishing since the 1930s to make way for cows, sheep, or plantations of conifers native to the United States that are a dead zone for local insects because none coevolved with the trees. The United Kingdom is now one of the least wooded countries in Europe, with just a fraction of its native woodlands remaining. An initiative by the British government to plant more trees has fallen woefully short of its targets, prompting conservationists to push for an "emergency tree plan" that would restore some of the lost habitat without falling back on commercial tree plantations.

But the absence of native woodland has now become so normalized that it usually goes unmentioned when there are periodic bouts of outrage over deforestation in the Amazon rainforest. Britain, a country stuffed with card-carrying conservationists obsessed with David Attenborough documentaries, somehow forgot that it exported its model of decimation to Brazil and elsewhere. "People say to me in Amazonia 'well you cut down all your old trees 2,000 years ago, why are you coming to tell me I can't?'" says Lees, who conducts regular research work in the Amazon. "We are the poor man of Europe for forest cover. Unless we get our own house in order it's hard to lecture other people about that."

When changes to rural areas are decried, what is invariably being referred to is farmland. This holds even for supposedly natural environments—the Peak District is around 90 percent privately owned farmland, after all. Britain, a densely populated country, gives over nearly three-quarters of its land to an agricultural system that has developed into a finely tuned machine. Technological advances and a drive for greater domestic food production has seen the United Kingdom's crop yields grow fourfold since the Second World War, with farms specializing with one or two crops and machinery making the work quicker and more efficient. Like elsewhere in the world, the traditional image of a cheerful family farmer tending to a small plot of land dotted with a few chickens has been eclipsed. As large agribusinesses have gained in strength, the number of British farms has fallen by two-thirds since the war.

This drive for efficiency has made Britain's countryside exceptionally tidy and regimented—fields have got larger, a few select varieties of wheat and barley have been favored, and half of all hedgerows, key habitats for pollinators and insect predators of crop pests, have disappeared in just a few generations. Around a tenth of the land is given over to the raising and killing of grouse by a small number of hunters, necessitating the extermination of any nearby animal that may possibly impinge upon a species that is treated like a living clay pigeon for shooters. Less sceptered isle and more sculpted isle, Britain has been manicured to death.

The ghosts of Britain's past natural abundance can be traced through "shadow woods," fragments of long-felled forests, and even via place names. There are more than 200 towns and villages in England named after wolves—Wolvesey, or "wolves island," is in Hampshire; Woolden, or "wolves' valley" is in Lancashire—despite the last wolf being snuffed out in the eighteenth century. Brown bears once lumbered around these isles, too, but by the early Middle Ages had also departed. Creatures such as the beaver—hunted to collapse for their meat, fur, and a secretion used in perfumes—and cranes—once so common that Henry II's chefs cooked up 115 of them at a Christmas feast in 1251 but then extinct in the country a few hundred years later—have been steadily reintroduced to pockets of Britain. But the subjugation of nature has never been fully reversed here.

The British countryside is verdantly beautiful, but it plays to a somewhat novel aesthetic. To the human eye, a neatly divided field of wheat or a lush swath of finely trimmed grass running right up to a stand of trees appears ordered, perhaps even attractive. But any satisfaction we may feel in the taming of our landscapes obscures the catastrophe that this inflicts upon insects, which thrive in what we would consider scruffy, disordered vegetation. Out of the chaos of untended grasslands and tangles of scrub comes a smorgasboard for a diverse array of insect life but also the annoyance of people who consider these areas unproductive eyesores. "People say you need to plant this or that for insects, but it turns out the plants were around us all along," says Lees. "People see scrub as messy and want lovely grassy areas but scrub is critically important for insects."

The untamed chalk grasslands that were once common across northwest Europe support a mix of herbs, flowers, and grasses that are the perfect home for many ground beetles and rarer insects, such as the phantom hoverfly, wart-biter bush cricket, and silver-spotted skipper butterfly. But around 80 percent of the United Kingdom's chalk grasslands have been turned into housing estates or pens for sheep. There's been a similar rate of loss for the distinctive purple heather that is ideal forage for bumblebees and, scientists have discovered, contains properties in the nectar that act as a natural medicine against parasites that plague the bees.

Diversity in cropland is already low and could get worse. A "three-crop rule" introduced by the European Union means that any farmer cultivating over 30 hectares (74 acres) is required to grow three different crops. That this was considered "diverse" may raise eyebrows, but it may get even less now that the United Kingdom has left the European Union and farmers push for the rule to be relaxed. This tiny palette means for much of the year there is no crop at all growing, just barren soil. "We've put big areas of crop in and we've reduced the natural habitat, there's nothing for the bees and other insects so they starve," says Barbara Smith, an agricultural ecologist at Coventry University. "We've taken a complex system and simplified it, taken everything out apart from one crop. It's like if the only food available was chips. Chips for everybody even if you don't eat chips."

Some even argue that the purportedly "natural" countryside of modern agriculture would provide a better home for insects if it was bulldozed and replaced with houses and their gardens. In his book *The Accidental Countryside*, naturalist Stephen Moss writes how environmentalist Chris Baines put forward this suggestion. "This may sound glib, but he was being entirely serious," Moss writes. "Most arable fields are monocultural deserts, with virtually no wildlife, whereas Britain's gardens are often home to a suite of former woodland birds and other wild creatures."

It's a disheartening comparison for those who have fought against the tide of urban sprawl, but for scientists the farmland trends are clear. Over the past two decades, Smith has noticed fewer and fewer insects on the agricultural land she studies, witnessing whole insect families slipping away and then vanishing. Rove beetles seem to do particularly badly with the advance of modern farming, as do solitary bees. Not only are they starved in the main fields; they often cannot even eke out a life in the margins.

Smith authored a paper in early 2020 that looked at the unglamorous but important role played by weeds in arable farming. Her paper found that arable fields need 10 percent weed cover to support enough insects to fully function in the food chain, feeding nearby birds such as the gray partridge. Cleavers, a plant also known as sticky willy, will attach its burs to you if you walk through and is hated by farmers because it can drown out crops. Nut grass is another unwanted weed but is less detrimental to crops. This distinction is moot, however—both support insects and both are indiscriminately removed from fields.

British farmers are now paid to set aside strips of wild grassland next to fields, but they are rarely linked as genuine wildlife corridors, and researchers have found these strips tend to benefit common, more mobile species rather than static, rarer creatures. Still, it was never farmers' intention to exterminate bees and butterflies, and there's hope that out of this emerging debate a system arrives that can support both food production and wildlife.

In the meantime, though, insect communities continue to slide to an unknown fate. "We don't really know how close we are to a tipping point," says Smith. "But there has been a big decline in insects in agricultural habitats over time and I think that's down to simplification of crops, the way that we manage fields and the use of pesticides. That's the simple story."

The quieting of the countryside isn't just a British disease, however. Agriculture has left a heavy imprint across Europe, fueled by a system of common subsidies that has supported farmers but also encouraged them to rip up hedges, wildflowers, and tall grasses to make way for crops.

Plants laden with food for pollinators have vanished from the Zurich region of Switzerland over the past century, a botanic comparison found, while seminatural grasslands in Sweden are just 10 percent of what they once were. The continent's farmland birds have dwindled, nitrogen washes from chemical-laden fields to create globs of algae off coastlines, and wetlands have been ploughed over, releasing huge belches of planet-warming greenhouse gases into the atmosphere.

As a boy, Benoit Fontaine took field notes of birds in a quiet area an hour's drive from Paris and remembers sitting outside on summer evenings for family dinners, insects clustering around a lamp on the table. He barely ever sees the bird species jotted in his notes anymore, while the evenings are mostly uninterrupted by insect visitors. "Everything's changed, and I've seen that in my lifetime," he says.

Fontaine, now a conservation biologist at France's National Museum of Natural History, was able to quantify some of this loss in a 2018 paper that reported that more than a third of birds have vanished from French farmland over the previous three decades. Once-common birds such as the meadow pipit and skylark have been cut down, the researchers suggested, by the dearth of insects on monocultural croplands.

On the great plains of Poitou, Champagne, and Beauce, hedgerows, groves, and ponds have been re-formed as flat, uniform plots ideal for the tractor and combine, largely to support the growing of maize, an animal feed. It's a dispiriting scene for naturalists. "It's a scenic landscape, but it's a desert," says Fontaine. "The French countryside has become a desert and it's the same story over the western world because industrial agriculture follows the same rules."

The practice of farming isn't inherently bad—in fact it has even been adopted by insects themselves. Scientists have discovered that the pale giant oak aphid has, seemingly for thousands of years, been "farmed" by ants, which keep the aphids underground in harsh weather and then in summer, when sap rises from English oak trees, march them up the tree trunk. These tiny shepherds do this so that the aphids can supply them with honeydew, the sugary water they excrete. In return, the ants herd the aphids into cozy "barns" they have built on trees from mosses and lichens. If danger approaches, the ants pick up their aphid flock and scurry with them to safety.

Meanwhile, in South America, leaf-cutter ants have spent the last 15 million years gathering plant matter to fertilize vast underground farms of fungi, a process that, over eons, has prompted the fungi to produce sacs full of protein—a perfect snack for the ants. Human agriculture, at around 10,000 years old, can almost seem an inconsequential blip when measured against these timeless marvels.

Yet, as Rachel Carson reminded us: "Wild creatures, like men, must have a place to live." Our reshaping of the planet is happening at such a pace that even ecosystems as vast as the Amazon can collapse in just a few decades, scientists have found. Once tipping points have been reached, fisheries can quickly dwindle away, huge lakes dry up, and coral reefs turn a deathly white. Our world appears stable and unchanging, until it isn't.

The squashing of habitat risks inflicting its own point of crisis for insects—and for us humans. Over the past half century, the world's farmland has been mired in monochrome sameness at a time when demand for pollinators has soared. A 2019 study warned that while we are not on the brink of mass starvation, dependence upon single crop monocultures unfriendly to pollinators "increases a country's economic and food security vulnerability."

Even if we are able to stave off worldwide food shortages, we are still left with a nasty brew of problems caused by a pernicious land use system that slashes away at our shared wild world. Insects—silently, almost casually—are being tossed into the gaping maw of this regime.

Almost all wildlife is suffering in our biodiversity crisis, but there's evidence that small invertebrate predators such as ladybugs and spiders are the worst hit when a natural habitat is razed for agricultural, urban, or road-building purposes. Erosion and pollution are peeling away soil and the cleanliness of waterways, further shrinking the world around insects. Even ostensibly protected areas aren't safe—more than a third of the world's conserved land is under "intense human pressure" from settlements, grazing, roads, railways, and nighttime lighting, researchers have calculated.

Insects, historically Earth's great survivors, have managed to adapt and even thrive with previous human tinkering of the environment. For centuries across Europe, woodland was selectively cut, or coppiced, in patches to provide timber and charcoal for cooking fuel. This arrangement was ideal for butterflies, which enjoyed the shafts of sunshine through the broken canopy and the subsequent regrowth that provided food for caterpillars and nectar sources for adult butterflies.

Modern logging techniques brought a new sort of brutal productivity to forest clearing, though, while charcoal has been replaced as a fuel by coal and gas. The butterflies were beneficiaries and then victims of our shifting practices. "We're not modifying it any longer in a way that is suited to their requirement, and they're going again," says Chris Thomas, biologist at the University of York.

The consolation is that they can rebound quickly if detrimental practices change again. Butterflies, like other insects, are being assailed by a barrage of threats. But if we give them just a few gasps of breathing space, even these most delicate, fragile-seeming species can find a way to make it. "Insect populations are like logs of wood that are pushed under water," says Roel van Klink, the author of the meta-analysis that found that terrestrial insects are declining by 9 percent a decade. "They want to come up, while we keep pushing them further down. But we can reduce the pressure so they can rise again."

This respite can come in unusual circumstances. During the coronavirus pandemic, some local authorities stopped the cutting of roadside grass verges, causing them to spring back with a shock of color. These narrow mini meadows suddenly became a rare refuge for an assortment of wildflowers of idiosyncratic nomenclature—yellow rattle, wild carrot, meadow cranesbill, greater knapweed, white campion—and acted as a magnet for a riot of insects, birds, and bats. The drop in car traffic also lifted a veil of pollution from floral scents, allowing bees to forage more easily.

Increasingly it's the scruffy incidentals of our world—the embankment next to the din of a highway, the greenery poking through railway tracks, the overgrown plot where a house once stood—that provide refuge to insects. Our interference with the environment is such that one highly endangered insect can only find a safe haven within, bizarrely, a firing range for explosives.

With its muddy brown wings, the Saint Francis' satyr would at first pass seem an unremarkable butterfly but for its unusual living arrangements. The butterfly is one of the rarest in the world, with almost all of the few thousand remaining individuals living in the environs of an artillery firing range at Fort Bragg, a US military base in central North Carolina. To the shuddering background noise of 181-kilogram (400-pound) bombs crunching into the ground, the butterfly happily flits around the firing range. By some quirk, the brute force of the US military has done a better job of safeguarding the Saint Francis' satyr than any outside conservation program.

The butterfly is fussy in an era when humans are serving up the same bland ecological meal across much of the world. It likes a habitat that is disturbed, but only a bit. It needs minor floods, but not too much water. It needs a few licks of flame to burn away overgrown plants but not wildfires that would completely torch its food. The North Carolina landscape once provided this regime, but then people started to suppress fires, hack down forests, and alter the hydrology of the environment. Amid the charred remains of fired ordnance, the artillery range re-creates a version of natural fire regimes and is the last fragment of land where the Saint Francis' satyr can make a suitable home.

When the butterfly appears—just twice a year for three weeks each time—Nick Haddad, a butterfly expert at Michigan State University, rushes to Fort Bragg. Haddad is given special permission to enter the artillery ranges to count and study the species and is struck each time by the feeling that he is in some ways stepping back in time. The target site for the firing range is a "moonscape," Haddad says, but once you step back from the bull's-eye there are untouched tracts of savanna, woodland, and swamp. Rare birds and snakes live here, as well as unusual plants, such as Venus flytraps and pitcher plants.

The butterflies are found in a grassy wetland, accessed via an area thick with shrubs and a mass of vines overhead. Haddad, wearing high boots due to the threat posed by unseen cottonmouth snakes, has to trek through this terrain on thin planks so as not to destroy the delicate vegetation the butterfly feeds on. In special restoration areas, he attempts to replicate the work of beavers by installing inflatable rubber bladders that mimic dams and provide the small-scale flooding the butterfly needs.

North Carolina is fringed with sweeping beaches in the east and towering mountains to the west, but in Haddad's opinion, the artillery range offers some of the most beautiful views of nature in the state. Re-creating this idyll outside the firing range is virtually impossible, but he's hopeful that other pockets of habitat for the Saint Francis' satyr can be established. The height of ambition probably falls below full restoration, however, and for rare butterflies outside islands like Fort Bragg, the outlook is bleak. We have been remarkably successful transforming our environment at an ever-increasing pace. Lots of insects will be able to share these changing states with us, but plenty won't, and there's currently little to save most of them from their doom. "I think the prognosis is terrible, it's really a frightening decline," Haddad says plainly about at-risk butterflies. "I tend towards being an optimist. When I say that the prognosis is terrible, you gotta keep that in mind I'm an optimist."

If you left Haddad in his undisturbed North Carolina swamp and ventured across the United States through the agricultural heartland of the Midwest, across the Great Plains, and into the goliath that is California's fruit and nut growing operation, you would probably share pessimism over the prospects of at-risk species along the journey.

Vast fields of pristine corn and soybean, grown largely as feed for animals crammed into enormous semi-automated livestock sheds, stretch endlessly across Iowa through to Minnesota. Place a honeybee hive near the soybean fields here and it will fare well initially, but then malnourishment will set in, scientists have found. Monitored bees in this environment "all just crashed and burned at the end of the year," says Amy Toth, an entomologist at Iowa State University. Further west, the beekeeping heartlands of North and South Dakota are being winnowed away as the corn and soy spreads and grasslands retreat. Pass through Idaho and Washington and you'll start seeing ruler-straight rows of potatoes, many of them grown by the Simplot farming empire to supply McDonald's with its french fries. Swing down to California's Central Valley and it can be easy to forget what untamed nature looks like as you survey fields carpeted with almonds, cotton, and citrus.

Were a grasshopper or butterfly to make such an epic journey, it would encounter "very few field margins or places to rest and recharge with nectar or an edible herb," says Jeff Pettis, an entomologist formerly with the US government. Much like the first white settlers, the insects would need to hurry across the dry West to get to the coast. They would face a major extra hurdle now, though. "The prairies were full of wildlife and food and clean water back then," Pettis says of the frontier journeys. "Now they are fields of corn and soybeans and little else."

These fields are growing as the number of farmers shrinks—America has fewer farmers today than it did during its Civil War, despite having a population eleven times greater. This is a symptom of a relentless push toward economies of scale, consolidation, and automation. Three-quarters of US farmed cropland is now controlled by 12 percent of farms, with the median farm size more than doubling to 499 hectares (1,234 acres) over the past three decades. Among all of this farmland, of course, are people, many of them located in sprawling car-dominated suburbs with manicured lawns, with huge snaking highways conveying them and the goods intended for their purchase. "None of these places are designed to support insects, they are for human use only," says Doug Tallamy, the University of Delaware entomologist who regularly makes the cross-country drive from Pennsylvania to see his grandchildren in Oregon.

The hours spent driving past mammoth fields where fringing wildflowers and weeds have been eliminated right up to the road infuriate Tallamy. "What frustrates me is that it's totally unnecessary, you don't have to sacrifice an ounce of yield to put native wildflowers at the edges," he seethes. "In trying to be neat in farming we are devastating insect populations. It's going to come back and bite us."

Insects have been devastated by the way we've altered the world around us, physically but also chemically. The battery of pesticides now routinely applied to our landscapes has created a toxic miasma for insects that scientists have only recently begun to quantify. Pest control for crops has been around for almost as long as there have been crops—the Sumerians of ancient Mesopotamia used sulfur compounds to vanquish insects and mites, while the Romans developed early rudimentary treatments to kill off weeds. Over the past century, however, it has been the chemical industry that has shaped a whole new arsenal of deadly weapons against invaders that nibble away or choke crops.

The broad umbrella of pesticides includes fungicides, deployed to eliminate parasitic fungi or their spores. There are also herbicides, used to remove weeds and most famously embodied by glyphosate, sold worldwide as Roundup. The emergence of chemicals of ever-greater potency since the 1970s has handed an edge to farmers in their perennial struggle against pests; glyphosate, according to the Australian crop scientist Stephen Powles, is a "one in a 100-year discovery that is as important for reliable global food production as penicillin is for battling disease." Glyphosate has also been linked to cancers emerging in tens of thousands of people, creating such a burden of litigation for Bayer that the conglomerate announced in July 2021 it was removing the herbicide from lawn and garden products sold in the United States.

But as chemical treatments have waged an ever-fiercer war against aphids, knotweed, and other enemies, it has increasingly caught all sorts of other insects in the cross fire. Herbicide use has boomed since the 1990s with the introduction of "Roundup-ready" crops resistant to the chemical, allowing it to be liberally spread on fields to take out the weeds. This has allowed the chemical manufacturers to target both ends of the process: the Roundup-ready herbicide-resistant seeds, sold by Monsanto, are marketed as the best defense against Roundup, the leading herbicide sold by Monsanto. But it has also led to herbicide leaching into the environment with some unforeseen consequences; glyphosate, for example, is thought to disturb bees' gut bacteria, leaving them more vulnerable to disease.

The impact of fungicides, which target molds and mildews rather than insects, has also surprised researchers. There is a significant correlation between fungicide use and the loss of bees, with lab-based studies finding that fungicides can worsen outbreaks of *Nosema*, a parasite that attacks honeybees and weakens colonies.

But the most deadly weapons aimed at insects are, as the name would suggest, insecticides, a class that has at its apex a group of chemicals called neonicotinoids. Chemically similar to nicotine—the word *neonicotinoid* means "new nicotine-like insecticides"—this new generation of insecticides was forged by Bayer, the German pharmaceutical giant, and has spawned eight commercially available variants produced by a range of manufacturers.

Over the past three decades, neonicotinoids have proved enduringly popular for use on everything from lawns to cropland and are now the most widely used insecticides in the world. The benefits are obvious: not only are the chemicals utterly devastating to sap-feeding pests like aphids, as well as fleas, certain wood-boring pests, and unwanted beetles, they are also considered "systemic" pesticides. This means that the chemicals don't just sit on the surface of a plant; they instead are absorbed and swiftly move through their host's circulatory system, reaching down into the roots and spreading to the extremities of the leaves and other tissues. "Neonics" provide a sort of all-over force field to 140 different types of crop, allowing farmers to be sanguine that their harvest, and livelihood, will be protected from the ravages of insect interlopers without the need for repeated chemical spraying.

In recent years, the all-in-one nature of neonicotinoids has been taken to new extremes. A small amount of the insecticide is now routinely applied to the coating of seeds that are sold to landowners, pushing chemicals into the lifeblood of plants from their very first growth spurt. Since the turn of the millennium, this farming format has become the default, with sales of neonicotinoid-coated seeds tripling in the United States alone.

Neonicotinoids are now so embedded within the food production of around 120 countries that residues of the insecticide have been found in, among other things, spinach, onions, green beans, tomatoes, and even baby food. In the United States, neonics have been detected in drinking water in Iowa, stubbornly remaining even after treatment. And in the United Kingdom, they have laced the waters of the River Waveney, the waterway that forms the border between Norfolk and Suffolk. When hundreds of people across China had their urine tested in 2017, almost every single sample contained neonicotinoids.

Any latent unease we may have about biting into a neonicotinoid-tinged strawberry is often assuaged by what we feel are corrective lifestyle choices, such as opting for organic produce. But while we may be able to bat away a more fundamental rethink of how we make food, there are no such comforts for insects.

The spring of 2008 was a savage one for Europe's honeybees, with millions perishing in France, the Netherlands, and Italy. The losses cut deepest in Germany, where the government had to set up containers along the autobahns for beekeepers to dump their moribund hives. An investigation traced the bee deaths to the use of clothianidin, a neonicotinoid, to stamp out an outbreak of corn rootworm. The finding prompted Bayer, the maker of the chemical, to pay compensation to beekeepers—with no admission of guilt.

A decade later in Brazil, around 500 million bees died in just a few months, the piles of dead bodies riddled with fipronil, an insecticide banned by the European Union and considered a possible human carcinogen by the United States. Brazil is increasingly awash in agricultural chemicals; since Jair Bolsonaro assumed the presidency, the country has been approving synthetic pesticides and fertilizers, some of them highly toxic, at a rate of around one a day. "We should be reducing the use of this stuff but we are increasing it," says Filipe Franca, a Brazilian ecologist. "Brazil is going completely against what we should be doing to save insects."

For many insects, the neonicotinoid era has been a punishingly cruel one that rivals that of DDT, an insecticide that gained infamy via Rachel Carson's *Silent Spring* and is now almost universally banned. It is probably even worse—neonicotinoids have been calculated to be around 7,000 times more toxic to bees than DDT. According to Dave Goulson, a single teaspoon of imidacloprid is enough to kill as many honeybees as there are people in India.

Neonicotinoids, which are water soluble, routinely seep into the soil and enter streams and rivers, coming into contact with a variety of terrestrial and aquatic insects. The chemicals find their way into wildflowers and taint their nectar and pollen, which are then picked up by unsuspecting pollinators. By some estimates, only 5 percent of the chemical actually stays within the target crop plant itself. Neonicotinoids set to work by assaulting receptors in an insect's nerve synapse, causing uncontrollable shaking and paralysis. While this is easily enough to prove fatal to a small pest like an aphid, the toxin has also been linked to the demise of butterflies, mayflies, dragonflies, wild bees, midges, and other invertebrates such as earthworms.

If insects escape death, there's a decent chance they will suffer a form of brain damage. Bees are shrewd operators able to grasp abstract math and able to pull strings and rotate levers in return for food, but chronic exposure to typical loads of clothianidin has been linked to cognitive damage that may scramble their learning and memory functions. This impaired function can be measured in distance—bees blighted by imidacloprid, another common neonicotinoid, fly shorter distances and for less time than unaffected bees, a crucial distinction for a species that requires repeated productive journeys to survive.

Imidacloprid has been linked to blindness in flies and colony losses among honeybees, while a third neonicotinoid, thiamethoxam, has been fingered as a potential culprit for cutting the reproductive output of bumblebee queens by a quarter. Both honeybees and wild bees need to hunker down in winter, feeding on stored honey or entering a hibernation-like state, but neonicotinoids have been blamed for reducing the chances of emerging safely from this stasis. The lives of bees are so entwined with the chemicals that when samples of honey were taken from around the world, traces of neonicotinoids were found in three-quarters of them.

Researchers at Imperial College London were keen to find out how profoundly bees are affected by neonicotinoids entering their colonies so set about using the technological might of modern research tools to peer inside bees' brains. In a laboratory, the team established a setup where the small fuzzy workers of bumblebee colonies were able to walk to a feeder filled with a sucrose solution containing neonicotinoids. The workers brought this food back to the colony, where it was used to rear the larvae that would form the next generation of bees. Once the young emerged from their pupae, the researchers tested the learning ability of half of these workers after three days and the other half after twelve days. The young bees were tested in a procedure that involved them being decapitated, with their removed heads then scanned by a micro-CT machine. "It's not something you want to do for the sake of it but it's for the good of science," Richard Gill, one of the Imperial researchers, says of the beheadings.

These results were compared with the young from colonies that were fed no pesticides, as well as those fed the chemicals only once they had emerged from the pupae as adults. All the bees were then given a test to see if they could learn to associate a smell with a food reward. The comparison was striking. The bees exposed to neonics during their larval stage performed poorly on the food reward test, with the area of their brains associated with learning and memory abnormally shrunken. The bees weren't given further pesticides as adults, so researchers were surprised to find that the insects had the same deficiencies as 12-day-olds as they did as 3-day-olds, suggesting a permanent level of brain damage inflicted by exposure in the very earliest stages of each bee's existence.

The regulatory safeguards around pesticides are routinely based on how deadly they are to certain creatures, but Gill argues that this overlooks critical nonlethal damage that can occur during bee development, analogous to how alcohol and drugs taken during pregnancy can cause damage to an unborn child. The toll from this prolonged harm could be enormous. "With wild bee colonies you only find the ones that survive, so there's a possibility we could be losing colonies at a rate we don't know about," Gill says. "People talk about colony collapse disorder exclusively about honeybees but is it possible that bumblebees are also suffering from a similar colony collapse disorder without being detected? We just don't know."

The impact is potentially deep, but it is also broad. Airborne pollinators are being exposed to neonicotinoids by collecting pollen and nectar, but the multitudinous communities of insects that crawl, burrow, and scuttle at the bases of plants face similar risks. Soils soaked with neonicotinoids have caused ground-nesting bees to be exposed to fatal levels of clothianidin, while slugs have been turned into toxic repositories that then indirectly kill off the beetles that prey upon and ingest them.

The tendrils of neonicotinoids are reaching deep into the crevices of our environment, from life in the soils and freshwaters to even the skies. A study of migrating white-crowned sparrows in Canada discovered that the birds lost weight just hours after gobbling seeds spiked with imidacloprid, delaying their onward migration and potentially impacting their reproductive success. The chemical, even in tiny doses, made the sparrows lethargic and suppressed their appetite, a symptom that would be familiar to anyone who regularly smokes. This malaise is likely not just a problem for the sparrows—researchers in the Netherlands found that concentrations of imidacloprid beyond a certain level knocked down insect-eating bird populations by 3.5 percent a year, on average.

In southwestern Japan, Lake Shinji is a body of brackish water home to fish, clams, and waterfowl and famed for its dreamy sunsets. In the early 1990s, rice farmers near the lake started using imidacloprid, and before long, the populations of arthropods that support the food chain, such as crustaceans and zooplankton, started to drop. To the dismay of locals, next to vanish were the eel and smelt that had seen their own food source diminish. The commercial fishery suffered a collapse it has yet to recover from, as ever-greater levels of imidacloprid are deployed on the nearby fields. The drenched environment of rice paddies is conducive to ferrying chemicals from fields to waterways, but researchers speculate that the same phenomenon could also be playing out in dry plains of wheat or corn. The Japanese scientists who elucidated the link between the pesticides and the decline of Lake Shinji concluded their research paper by quoting

Silent Spring, where Rachel Carson laments the ability of pesticides to "still the song of birds and the leaping of fish in the streams."

"The world of systemic insecticides is a weird world, surpassing the imaginings of the brothers Grimm," Carson wrote. "It is a world where the enchanted forest of the fairy tales has become a poisonous forest." Six decades on, everything and yet nothing seems to have changed. "The ecological and economic impact of neonicotinoids on the inland waters of Japan confirms Carson's prophecy," the researchers wrote.

The entrenched use of neonicotinoids means their legacy is set to linger for some time. In the United States alone, almost every corn seed and cotton boll found in a field is treated with neonics. Around half of the seeds in soybeans are, too. In all, neonicotinoids typically cover around 61 million hectares (150 million acres) of American cropland, an area about the size of Texas.

The chemicals tend to accumulate rather than wash through fields, slowly layering new levels of toxicity. The farmland of countries with heavy insecticide use is probably more laden with lethal or deleterious chemicals than at any point in history. According to one study, the past quarter of a century has seen US agriculture become a whopping forty-eight times more toxic to insect life, with neonicotinoids responsible for almost all of this noxious surge. Across vast, featureless fields, insects are being systematically maimed, befuddled, and exterminated. "Our insects are now playing in a dirty playground and they just don't have the diversity or genetic makeup to withstand this," says Alex Zomchek, an apiculturist at Miami University.

Even as the sheer weight of insecticides doled out onto fields has fallen in recent decades across much of the United States, the hazard for insects keeps escalating. One analysis found that the region defined as the American heartland—Iowa, Illinois, Indiana, most of Missouri, and parts of five other states—has become an incredible 121 times more toxic over the past twenty years for bees. What may look like a benign field of corn to us is, to an insect, more akin to a home that has been replaced by a fetid pit filled with whirring buzzsaws and famished crocodiles. "You add a pesticide to the system, but often there are still pesticides left from the previous year," says Christina Grozinger, entomologist at Pennsylvania State University and author of the analysis. "The neonics and other chemicals that linger, the concentrations keep growing. It's just going to build and build and build."

The impact of insecticides is so severe that it can spill over from target areas into places of natural abundance, as is suspected in the countryside of Krefeld, in Germany, where the declines of insects were recorded in a sort of chessboard landscape where agriculture abuts sanctuary. A varied landscape can offset some of the harm caused by pesticides, too, with a study of honeybees and wild bees near canola fields in the United Kingdom, Germany, and Hungary showing that bees able to mix up their diet with visits to nearby natural areas fared better than those surrounded only by a pesticide-laden monoculture.

But it's not just about bees. These eusocial creatures are studied endlessly due to their charisma and importance to our lives, but Grozinger is in no doubt that whole strata of overlooked insects are being pummeled. Many of them aren't able to escape to islands of refuge amid fields that unfurl like carpets to the horizon. "There's this beautiful diversity of insects that we know very little about and the vast majority of them in a situation like this are not going to do well," she says. "There's going to be a handful that will be fine."

The frustration for Grozinger and other entomologists, beyond the loss and derangement of so many insects, is that it may well all be for nothing. Penn State University has an instructive fact sheet that illustrates the bumbling wastefulness: a nascent soybean plant is saturated with neonicotinoids in its early development, but it's not until midsummer, when the plant is more developed and the chemicals have largely seeped away into the environment, that the target pest, the aphid, arrives in numbers. The peak of the pesticide doesn't match the peak of the pest.

Several dozen scientists looked at this issue more closely in 2019 and made a rather dispiriting finding: from nearly 200 studies of soybean crops across the US Midwest, there was little evidence to suggest neonicotinoids improved the harvest. The chemicals provide "negligible benefits to US farmers" when the cost of the treatments is factored in, the researchers found. A catalyst of this shortfall is, perversely, the things neonicotinoids do manage to eliminate. By taking out a sizable chunk of predatory insects, the chemicals can counterproductively aid pests such as aphids, black cutworm, and armyworm to freely munch away on the crops.

A study of nearly 1,000 farms of all types across France found that 94 percent would lose no production if they reduced pesticides, with a sizable minority actually able to churn out more food and fiber with a lighter chemical load. The situation with insecticides is particularly striking—nearly nine in ten of the farms would increase their production with lower levels of these chemicals, with no farms at all experiencing a drop in output.

The research, published in 2017, came shortly after a UN report waded into claims made by pesticide manufacturers that their product was vital to help feed a global population expected to balloon to more than 9 billion people by 2050. The report maintained that this belief was a "myth" and that chemical companies are in denial over the "catastrophic impacts on the environment, human health and society as a whole" caused by pesticides.

The backlash against the use of pesticides has fueled a small but growing movement within agriculture to shift away from chemicals almost entirely. "Pesticides are absolutely unnecessary," says Jon Lundgren, an entomologist who claims that his own South Dakota farm is plagued by fewer pests by embracing the principles of regenerative agriculture, whereby soils are never left bare, biodiversity is supported so that insect predators act like nightclub bouncers for the crops, and the farm itself is a mishmash of livestock, crops, and orchards rather than uniform plains. "In the current system, the natural resource base is crashing," Lundgren says. "The insect apocalypse is just the first sign of this."

Farmers would, ideally, revert to techniques such as rotating crops, carefully managing sowing dates, and doing plenty of weeding with machines rather than sprays. But even scientists who have discovered that crops can still flourish without chemicals point out that this doesn't necessarily mean the treatments are completely useless, more that they are being overused in an indiscriminate and destructive manner. "If this was such a great pesticide you would be able to use less of it and maybe see flat levels of total toxicity," says Grozinger of neonics. "You wouldn't see the increase we are seeing, which is problematic. It suggests it's not in response to an actual issue, like pest control, but some other factor."

Farmers have been caught in a "circular addiction," Lundgren says, whereby the use of insecticides has led to continued, and intensified, use of chemicals to deal with the consequences of the initial biodiversity loss. But this cycle is never inevitable. Previous generations of farmers managed to reap a decent bounty without bombarding their crops with a cocktail of poisons, so why not now?

Part of the answer lies in the sheer power of agribusiness—the traditional "big six" has morphed through mergers in recent years into an even bigger three: Bayer-Monsanto, Dow-DuPont, and Syngenta-ChemChina—to promote the use of insecticides as imperatives to farmers, who sometimes aren't entirely sure what exactly is in the coating of the seeds they are being sold en masse, as well as to deter lawmakers and regulators from stamping out their use.

Jeff Pettis has seen the clout of industry at close quarters. During his long career as a scientist at the US Department of Agriculture (USDA), Pettis sought to find out what impact neonics were having on honeybees, so he started feeding colonies with protein patties, essentially miniature hamburgers for bees, containing imidacloprid. The amounts added were minute, at least ten times less than the safe thresholds advised by Bayer. "It was the equivalent of putting five drops in an Olympic-sized swimming pool and mixing it equally," Pettis says. "We are talking very small amounts." After a few months, the researchers could see that new, young bees fed the affected protein were significantly more likely to be struck down by *Nosema*, the fungal gut parasite. This impact, the researchers wrote, suggested that pesticides could be a "major contributor" to increased mortality of bees, including colony collapse disorder, the disastrous phenomenon where bees suddenly abandon a hive.

The pesticide makers then embarked on a campaign that Pettis likens to the tobacco industry's disparagement of science linking smoking to various cancers. His work was criticized for being unrealistic as to what would happen in a field or for attempting to wind the clock back to the bad old days of repeated mass spraying of crops. He noticed that his department was restricting his ability to talk to the press about his findings or to hold public meetings on the topic. He was upbraided by a Republican member of Congress for not "sticking to the script" by talking about neonics. Pettis was demoted; eventually he resigned. "Wherever they could they would cast doubt," Pettis says. "They were defending the status quo, I suppose."

Pesticide manufacturers have funded groups that dispute research finding neonics are harmful, co-opted previously critical scientists, and backed bee health initiatives that are heavily skewed toward targeting mites rather than chemicals. Emails show that Monsanto, which is now part of Bayer, attempted to orchestrate a campaign to discredit scientists linking Roundup to cancer concerns, while Bayer has created online videos depicting people worried about pesticides as conspiracy theorists fond of talking to flowers and downplaying the harm caused by the chemicals. "The truth is, our body handles all sorts of chemicals every day, which is normal," a voiceover states over footage of a sugar cube splashing into a cup of tea and a woman applying lipstick.

No one can be quite sure if any of these efforts had a role in subverting lawmakers or triggering Pettis's demotion, but the experience was chastening. "When I lost the trust of USDA and my ability to speak to bee diseases and problems, that was just an uncomfortable situation and I chose to resign," Pettis says. "In hindsight, there was a lot of pressure. The pressure was coming in a lot of different ways. It was clear the pesticide industry wasn't happy with what I had to say."

Bayer, for its part, points out that honeybee colony numbers have grown worldwide over the past half century—the company tends not to mention the trends among wild bees—and that an extra 1.2 million hectares (3 million acres) of farmland would be needed to match the yields provided by neonic seed treatments. Certainly, the situation is nuanced, and many different calamities are hurting bees at the same time, not all stemming from one or two multinational corporations. But the continuing use of the most hazardous pesticides, linked by independent experts to damage in humans and the wider environment, stubbornly remains a financial necessity for the industry.

Data produced by Phillips McDougall, the leading agribusiness analysts, show that the five largest pesticide makers sold $4.8 billion in highly hazardous pesticides in 2018, making up more than a third of their total income. Around 10 percent of all sales generated by the manufacturers came from pesticides judged toxic to bees. Maintaining these sales relies, in significant measure, on politicians and the public not being spooked by bothersome scientists or campaign groups.

It was a major blow to the pesticide makers, then, when the European Union opted in 2018 to ban all outdoor uses of clothianidin, imidacloprid, and thiamethoxam, the three most common neonicotinoids. The bloc had previously restricted the use of the chemicals on flowering crops that attract bees, such as oilseed rape, but decided to impose a more sweeping ban after an assessment found that the chemicals still posed a major risk to bees, as well as to the health of soil and waterways. The crackdown was the first major regulatory assault on the insect crisis since the issue exploded in the public realm and was met with jubilation among campaigners. "Authorizing neonicotinoids a quarter of a century ago was a mistake and led to an environmental disaster," said a triumphant Martin Dermine at Pesticide Action Network Europe on the day of the decision. "Today's vote is historic."

Some countries within the European Union went even further. France decided to ban the neonicotinoids thiacloprid and acetamiprid on top of the trio outlawed by the European Union, extending the rule to greenhouses as well as outdoor uses. Meanwhile, Austria, the Czech Republic, Italy, and the Netherlands have moved to restrict the use of the weed killer glyphosate. Germany, home of

Bayer, the ultimate owner of Roundup, announced plans to do likewise by 2023. "What harms insects also harms people," opined German environment minister Svenja Schulze. "What we need is more humming and buzzing."

It's tempting to think of these bans as a panacea that will see a resurgence in European insect life, but the reality is bogged down in complexities. Neonics appear to linger a long time in the environment—the previous restrictions still left behind traces of the chemicals in bees and nectar for several years—and there is no certainty that the next batch of tools used against crop pests won't be similarly harmful.

Dave Goulson could be forgiven for taking a victory lap after corralling more than 240 other scientists to sign an open letter calling for the international banning of neonicotinoids, but his outlook is more phlegmatic. The previous triumph over DDT was meant to have acted as a capstone to Carson's legacy and ease the pressure on the natural world, and yet insects have never faced more crushing circumstances than now. Withdrawing a few weapons doesn't alter the war if the battle strategies remain the same. "We do seem to keep making the same mistake over and over again," Goulson says.

Researchers have amassed three decades of evidence on the impact of neonicotinoids, a body of work used by the European Union to implement its ban. The replacement class of insecticides, when it emerges, will reset the clock on the investigations. A pattern of pressure and release is fated to repeat itself as long as the mold of modern agriculture remains unbroken. "It's an endless cycle where you ban something and it's replaced by something else and twenty years later you realize that that's also harmful to the environment," says Goulson. "So you ban that and on we go."

Outside Europe, the cycle is taking longer to spin. Pesticide makers have high ambitions to greatly boost sales in Africa, a continent that has already shown signs, in Ghana, that widespread use of neonics in cocoa cultivation harms the plants' natural pollinator, a midge. The chemicals have also cut down the natural enemies of certain cocoa plant pests that subsequently blossomed in number—an anxiety-ridden trend for those of us who enjoy chocolate.

In the United States, some initial maneuvers to restrict neonicotinoids during Barack Obama's presidency fizzled out with Donald Trump's reversal of even the most meager of gestures, such as a ban on using neonics within federal nature reserves. Beekeepers, increasingly stretched in providing the enormous amount of pollination required to grease the wheels of American agriculture, have resorted to suing in court to force a crackdown on neonics but have been repeatedly thwarted. Even if they prove successful, the reprieve will likely be temporary, and not just because a fresh pool of insecticides will be cooked up in a lab to replace neonics. Without other changes to the way we use land, simply eliminating pesticides could end up causing more harm to insects in the scramble to feed a growing global population.

While many farms could radically slash pesticide use and fare just as well, some studies have shown that the yields of certain unprotected crops are at risk of being decimated by pests. If there was no immediate replacement for pesticides, farms would likely drastically increase the amount of growing land to make up for the shortfall in harvests. Given the projected growth in worldwide population, the FAO estimates, based on current dietary trends, that we'll need to raise an extra 200 million metric tons (220 million US tons) of meat and a billion extra tons of cereal crops a year until 2050 to keep pace.

The land sacrificed for this extra food would trigger disaster for insects as well as the wider environment. "If we didn't add the pesticides, we would need to have half as much land, again, under cultivation," says Chris Thomas, the University of York biologist. "That's a staggering thought. Almost all of the remaining productive soil that could produce good crops and livestock is under tropical rainforest right now."

Deeper, more fundamental reforms would have to take place to address the insect crisis without making such dire trade-offs. A philosophy of integrated pest management, where pesticides are used as a last resort after a range of approaches such as crop rotation and encouraging natural predators of pests, would have to take hold, perhaps alongside the embrace of vertical farming, where produce is grown in towering stacks in futuristic, soilless indoor settings with water and nutrients delivered via hydroponics.

The daily chores undertaken in our own homes may seem far removed from the industrial overhaul needed to ease pressure on insects. Not many of us, after all, sow toxin-covered seeds in fields or raze a wildflower meadow to the ground as part of our regular routines. But we are complicit in the insect crisis, too, both indirectly through the choices we make around how we consume resources and directly through our actions at home. Since the 1950s, households have been routinely stocked with an array of insect sprays to eliminate cockroaches, ants, flies, and other flying and scurrying undesirables found around the home and garden. This blitz has not only bred a strain of chemical resistance in mosquitoes, which need ever-evolving treatments in places where malaria and dengue are rife, but also created an agricultural situation in miniature where pest predators are caught up in the collateral damage, allowing space for pests to proliferate.

A seemingly benign and pleasing symbol of Western household affluence has also proved a major foe to insect life. Lawns have become a must-have adornment to suburban life in many countries, all to the same uniform aesthetic—verdantly green, weed free, neatly trimmed, and lush underfoot, as if they are a modified extension of indoor carpeting.

In countries such as China, lawns are the domain of public parks for viewing rather than recreation, but in Europe, Australia, and the United States, they are now strongly associated with the pride of homeownership and a certain quality of diligent labor. "In the US more than in any other country, the lawn is a mostly symbolic element of prestige and status," says Maria Ignatieva, a landscape architect at the University of Western Australia who has spent years studying lawns around the world.

An enormous $36 billion industry has developed around herbicides and mowers to keep American lawns to this desired aesthetic. Residential lawns are doused in an estimated 26.5 billion liters (7 billion gallons) of water each day and strafed by 27 million kilograms (59 million pounds) of pesticides a year. Plants useful for insects do sprout in this otherwise sterile environment, but are often deemed weeds and quickly snuffed out by assiduous homeowners. "If it is just a manicured, mowed lawn many pollinator insects have not too many options to find flowers," says Ignatieva.

Lawns may seem like a minor element of the environment, but in aggregate they dominate the green spaces free of agriculture, urban concrete, and industry in many countries. In 2005, scientists at NASA used satellite imagery to make a startling discovery: lawns, including commercial lawns and golf courses, are the single largest irrigated crop in America in terms of surface area. Lawns sprawl across 128,000 square kilometers (49,421 square miles) of the United States, about three times the area taken up by corn crops.

We have constructed deceitfully comforting edifices of beauty and order around us. Neatly divided fields of crops, lush green swards of grass, and gorgeously exotic ornamental plants provide us an illusion of vibrant abundance, but it's worth pausing for a moment to note what's missing. Our surroundings should be rife with insects, along with the birds and other creatures that depend on them. Instead, we've done everything possible to banish them.

Some fairly minor changes could break up the monotony of our domestic environment and boost insect life. Incorporating wilder plants such as clovers and thyme can provide vital stepping stones for insects. Letting the grass grow a little allows more variety in, and therefore more life. Creatures such as earwigs, beetles, and spiders like to loiter under accumulated leaves, so we could reconsider how often we rake the yard. Impromptu leaf homes are destroyed so regularly that the German government recently urged its citizens to avoid using leaf blowers because they are "fatal to insects in the foliage."

In the United Kingdom, the British Ecological Society has pleaded with people to spare the humble dandelion, a valuable food source for pollinators such as solitary bees, honeybees, and hoverflies, and to avoid instinctive favorites such as the archetypal rose, which contains little nectar or pollen. Letting a little chaos reign amid the onions and carrots in the vegetable patch wouldn't hurt, either. "This whole business of keeping your lawn clipped and pulling the weeds out is part of some British obsession with tidiness," professor Jane Memmott, president of the society, told *The Guardian*. People should be aiming for a certain "bohemian untidiness," as Memmott delightfully put it. "You can't personally help tigers, whales, and elephants but you really can do something for the insects, birds, and plants that are local to you," she added.

The cult of the lawn isn't likely to vanish, but Douglas Tallamy hopes it can at least evolve. "It's mindless to have this postcard perfect, green lawn," says Tallamy, who has calculated that if American homeowners converted half their land to native plants, they would create more habitat for insects than all of the national parks in the lower forty-eight states combined—Yellowstone, Yosemite, the Everglades, the lot. This would help open up connected wildlife corridors, allowing insects to move safely through desolate, human-dominated environs.

Like Memmott, Tallamy believes that even a slight reorientation could provide insects with a major reprieve. "I'm not saying put a meadow in your front yard, that would be culturally too shocking right now," Tallamy says. "We can still manicure the lawn we have and show that we're good citizens, but we need less of that signal. We don't need to do that with acres and acres. We can't afford to do it, there's nothing good about it."

As if spreading nerve agents into insects' domain isn't bad enough, we've also taken it upon ourselves to befuddle them by illuminating the night. Artificial light has been a problem for insects since the light bulb arrived in the nineteenth century, but the more recent spread of human lighting, from streetlights to sporting stadiums to flaring gas in oil fields, not to mention the arrival of dazzling LED, means that light pollution now affects around a quarter of the globe's land surface. This problem immediately conjures up a mental image of a moth ceaselessly circling a bare light bulb, having confused it for the moon. Around a third of moths that get trapped in this fruitless orbit die by the morning, either from exhaustion or being eaten by a predator.

Moths are far from the only victims, however, with a recent research paper warning that light pollution is an "important—but often overlooked—bringer of the insect apocalypse." Mayflies, which only live for a day, seek out polarized light and often mistakenly lay their eggs on roads or other perilous places. Light can harm the development of young insects and disrupt the feeding ability of creatures, such as stick insects, that actively avoid bright places. Billions of flying insects smash into car headlights, while insects such as the corn earworm stop mating if light levels are above that of a quarter moon.

Flooding the night with light can even hamper the pollination of plants, an activity usually associated with balmy, sunny days. Nocturnal visits to plants by moths, beetles, and bugs are stymied so much by nighttime light that a study by Swiss researchers found that fruit production can fall by 13 percent as a result, even if there are plenty of daytime insect visitors. This drop in night shift pollination due to distracting lights moved the United Nations to warn of "deeply concerning" consequences for food security.

As is common in the insect crisis, the link between action and consequence can seem a little blurred. Flicking on the porch light doesn't immediately provoke a wave of guilt over food shortages and scores of tormented creatures. But light pollution is one of the most intractable problems for insects; while some can adapt to habitat loss or a warming planet, the division between night and day has been hardwired into every insect since the dawn of evolutionary time. There is no escape from it.

Distressingly, excess nighttime light is menacing one of the most magical insects we ever see—fireflies. They are also referred to as glowworms in Europe and lightning bugs in North America. They are in fact not flies, nor worms, nor bugs, but beetles that use bioluminescence to emit light.

Growing up in South Texas, Ben Pfeiffer would often witness a carnival of flickering firefly light at night on the family ranch. He could fill jars with the insects, which gave off an array of different light patterns. "When you get five different species flashing different lights all around you, it's amazing," he says. Filling a jar would be trickier these days. "People usually just see one light pattern now. Once the declines happened, the intensity and brilliance diminished."

Firefly researchers around the world have reported that these most radiant of insects are in trouble for many of the reasons other insects are in trouble. Fireflies often choose to live on the sodden edge of riverbanks, laying their eggs in mud and feeding on snails and slugs. Plonk a housing development in these riparian areas and whole species can get scratched out.

Pfeiffer frets about carrizo cane, an invasive weed that is clogging up firefly habitat all along the Rio Grande, as well as the general levels of pollution in rivers and springs that harm the species. But he is also irked by light pollution he sees radiating into the night. Even in many quieter places, it is never truly dark, with a diffuse illumination from lights creating an effect known as "skyglow." This light intrudes upon fireflies' ability to signal to each other. The insects use showy patterns to find a mate, with females picking males that can display the brightest, quickest flashes. Many firefly species need complete darkness to relay these messages, so light pollution can pose a chronic barrier to the cycle of reproduction.

Modern lighting technology is particularly problematic. A British research team has studied the eyes of male common glowworms, which are attracted to glowing females. The males could detect the green light emitted by the females, but when blue light was added, they struggled to locate the females. New LED streetlights that give off a bluish tinge are therefore more likely to disrupt the fireflies than old-fashioned sodium streetlights.

Worst of all, Pfeiffer says, is the very harsh white LED light that can be uncomfortable to look directly at. "If it hurts your eyes, what do you think it's doing to a firefly?" he says. Increasingly, one of the only firefly species able to adapt to the incandescent world we've created is the big dipper, a creature found as far north as New York that is happy to do its courtship around sunset, so isn't too bothered by the extra nighttime light. Pfeiffer, now a firefly researcher, was once surprised to see a big dipper in an oak tree at a busy intersection outside a Walmart, the insect giving off a signature flashing pattern that resembles a reverse J. Such occasions become more memorable the rarer they become.

The lack of firefly sightings is something that people now regularly mention to Pfeiffer, a jolt of dawning recognition similar to that evoked by Anders Møller's clear car windshields, thousands of miles away in Denmark. "People are waking up to this," he says. "I can see the changes happening before my eyes. Pollution is there in various forms, from light to trash to pesticides. I think you'll have a massive collapse of insect diversity in a wide variety of areas and it's going to be shocking to people."

5

In the Teeth of the Climate Emergency

Climate change is warping the settled order of life at such unappreciated speed that it is making our nomenclature redundant and even absurd.

The Monarch Butterfly Biosphere Reserve in central Mexico will, in the foreseeable future, support no monarch butterflies. In the Pacific Ocean, the nation of Tuvalu, a name that means "eight standing together" in reference to each of its inhabited islands, already has two islands on the verge of being utterly swallowed up by sea level rise and erosion. The remainder are set to follow. Startling transformations are occurring in chillier climes, too. Glacier National Park, a largely pristine wilderness in the northern extremities of Montana, is named after its spectacular glacier-carved mountains and valleys, but soon this title will be rather incongruous.

Of its 150 glaciers present in the mid-nineteenth century, just 25 now remain. These, too, will recede to nothingness—substantial glaciers could melt away as quickly as the year 2030, although some will hang on until the end of the century. "The glaciers are disappearing rapidly," says Clint Muhlfield, a research aquatic ecologist for the US Geological Survey stationed in the park. Muhlfield has contributed to research showing that the area covered by glaciers here has shrunk by 73 percent over the past 170 years. "The Glacier National Park will eventually have no glaciers," he says. "That will be a huge change."

In virtually every aspect, this area is an untouched refuge. All of the animals and plant species present when Lewis and Clark's westward expedition arrived in the region in 1806 are still here—lakes and streams are thick with cutthroat trout and sockeye salmon, bald eagles and osprey swoop overhead, grizzly bears, moose, and wolverines pad the precipitous terrain. This setting provides "the best care-killing scenery on the continent," according to John Muir, the Scottish American naturalist, who marveled at the lakes of glacier water, waterfalls, forests, and nemophila-blue skies. "Give a month at least to this precious reserve," Muir wrote in 1901. "The time will not be taken from the sum of your life. Instead of shortening it, it will indefinitely lengthen it and make you truly immortal."

The national park, the narrowest point in the Rocky Mountain caldera, is the shimmering jewel in a broader ecosystem known as the Crown of the Continent, which stretches from Montana into the Canadian provinces of Alberta and British Columbia. This area contains such a rich network of intact waterways that it can carry a droplet of water to almost all points of the compass: east to the Mississippi and the Atlantic Ocean, north to the Arctic Ocean, or westward to the Pacific.

But in its unsentimental relentlessness, the cudgel of climate change doesn't mercifully bypass places of beauty or even ones we have heavily protected. "Climate change has essentially impacted every square centimeter on the face of Earth at this point," Muhlfield says. Rather than being somehow shielded from rising temperatures, Glacier National Park is in fact being slowly roasted, heating up at around two to three times the global average. The escalating heat is shrinking the glaciers, while changes in precipitation mean that instead of replenishing snowfall, the park is getting more rain. This has increased the frequency of autumnal and winter flooding events, but conversely also means that less water is running off the reduced snowpack as it melts each year. This gushing meltwater usually means that the heaviest stream flows occur in spring, but this high point is now at least two weeks earlier, on average, than it was in the 1950s because there is less ice to feed the waterways.

This is a potentially terminal situation for two species of stone fly—the western glacier stone fly and the meltwater lednian stone fly—that live in these environs and nowhere else. Both creatures are brown in color, have two sets of translucent wings, barely reach 1 centimeter (less than half an inch) in body length, and base their existence around cold, clean streams flowing from glaciers. The aquatic insects spend their whole life cycles in short sections of alpine streams directly below glaciers, from egg to nymph to adults that in turn lay eggs. The male and female western glacier stone flies communicate by drumming their abdomens on the pebbles and other material found at the bottom of the streams.

Such inconspicuous insects are rarely compared to the most famed beasts of our world, but Muhlfield likes to call these stone flies the "polar bears of Glacier National Park." In 2019, menaced by the loss of the frigid runoff from glaciers, the two insect species became the first animals aside from the polar bear to be listed as endangered by the United States due to the threat of climate change. Species better suited to warmer waters are able to creep up the mountainside as things heat up, but the western glacier stone fly and meltwater lednian stone fly are boxed in. "They are literally at the top of these mountains and there's nowhere left for them to go," says Muhlfield. "They have to shift or go extinct and they are running out of space pretty quickly. It's like a squeeze-play at the top of the continent."

The stone flies would be just another couple of obscure insect species to blink out sight unseen were it not for researchers such as Muhlfield who have to hike, sometimes for several days, to remote streams to find the insects. The researchers typically go in summer, when Glacier is resplendent in wildflowers, to avoid being impeded by snow. The journey, however, still involves plenty of bushwhacking far from recognized trails.

Once at the waterways, samples are taken using a funnel-like device, which looks a little like a wind sock, that is placed vertically into the substrate of the streambed to capture what's there. The process, dutifully repeated hundreds of times, has helped reveal a picture where the stone flies are being increasingly backed into a corner of the few streams that remain cold enough for them. Climate change's toll on insects will only grow, even if it will be far less visible to us than a miserable polar bear on an iceless sea. "In some cases it might be too late," says Muhlfield. "We really don't know what we're losing."

The implications of climate change are terrifying. They are also frustratingly nebulous, even as we are assaulted with a growing ferocity that is razing entire towns in wildfires, submerging others under the rising seas, and baking millions in punishing heat waves. We don't know quite how bad it will get or if the paths to a more moderate version of the crisis have been completely shut off yet, although almost all the evidence suggests we have left it too late to avoid suffering on a huge scale. One recent study found that even under the most optimistic scenario, within fifty years 1.2 billion people will live in temperatures currently found only in the most scorching parts of the Sahara. Part of why this can seem abstract to us is not only the slow-motion nature of the disaster but also its unprecedented awfulness.

Humans will not be the only species to suffer from the calamity caused by industry pumping huge volumes of planet-warming gases into the atmosphere. Huge waves of die-offs will be triggered across the animal kingdom as coral reefs turn ghostly white and tropical rainforests collapse. For a period, some researchers suspected that insects may be less affected, or at least more adaptable, than mammals, birds, and other groups of creatures. With their large, elastic populations and their defiance of previous mass extinction events, surely insects will do better than most in the teeth of the climate emergency?

One of the first studies to look at this properly, in 2018, turned this assumption on its head. Researchers gathered data on current geographic ranges and current climate conditions of 115,000 species across the animal and plant world in order to find out what combinations of temperature, rainfall, and other climatic conditions each species can tolerate. They used computer models to project how changes in climate would then alter the geographic range of each species under different levels of global warming, from 1.5°C (2.7°F) above the pre-industrial era to 3.2°C (5.8°F) above.

Rachel Warren, the University of East Anglia biologist who led this work, describes this process as a bit like looking down from space and seeing how certain animals are bound within discrete slices of the world. Heat things up a little bit and an animal can move a little closer to the poles or up a mountain to find suitable cool temperatures. But there's a limit to this. Crank the temperature up further or faster and animals struggle to survive because they can't keep up or because it is now too hot even on the mountaintop. At some point, there's nowhere else to go.

Separate studies have shown that creatures from fish to primates are being constricted by this phenomenon, but surprisingly, insects are faring worst of all, according to Warren's research. At 3.2°C of warming, which the world is on track for by the end of the century in the absence of major emissions reductions, half of all insect species will lose more than half of their current habitable range. This is around double the proportion of vertebrates and higher even than for plants, which lack wings or legs to quickly relocate themselves. This huge contraction in livable space is being heaped onto the existing woes faced by insects from habitat loss and pesticide use, threats not considered by the study. "The insects that are still hanging in there are then going to get hit by climate change as well," Warren says. "That basically means that things are a lot worse than these numbers say."

While the research factors in species' ability to migrate to more hospitable climes, it doesn't consider the impact of severed interactions between dependent species nor the damage from extreme weather events stirred by climate change. Many potential climate refuges have been turned into agricultural or industrial plots that are hostile to insects, too. These compounding elements mean that the devastation across insect species is hard to quantify.

Some insects, such as dragonflies, are nimble enough to cope with the creeping change. Unfortunately, most are not. Butterflies and moths are also often quite mobile, but in different stages of their life cycle they rely on certain terrestrial conditions and particular plant foods, and so many are still acutely vulnerable. Pollinators such as bees and flies can generally move only short distances, exacerbating an emerging food security crisis where farmers will struggle to grow certain foods not just due to a lack of pollination but because, beyond an increase of 3°C or so, vast swaths of land simply become unsuitable for many crops. The area available to grow abundant coffee and chocolate, for example, is expected to shrivel as tropical regions surge to temperatures unseen in human history.

"Insects are fundamental to our ecosystem and we do fear that there could be ecosystem collapse, basically, if nothing is done," Warren says, explaining that with a 4°C (7.2°F) increase in warming there would be a spiraling risk of famine, particularly in countries that cannot afford substantial imports of food from overseas. "It's not a pretty picture."

The climate crisis interlocks with so many other maladies—poverty, racism, social unrest, inequality, the crushing of biodiversity—that it can be easy to overlook how it has viciously ensnared insects. The problem also feels more intractable; bans can be placed on insecticides, while farmland and our cities can be made more conducive to insect life, but there is no escape from the upheaval of climate change. "Climate change is tricky because it's hard to combat," says Matt Forister, a professor of biology at the University of Nevada. "Pesticides are relatively straightforward by comparison but climate change can alter the water table, affect the predators, affect the plants. It's multifaceted."

Forister spends much of his fieldwork in the sweeping expanses of Nevada and northern California and is constantly struck at how butterflies, bees, and other insects are struggling when they have so much space to inhabit. He found himself pondering: "Why is it that these little things are suffering so much?" But a closer look made it clear that the river habitats are changing, the meadows are degraded, the wetlands are gone in many places, and even what Forister refers to as the "scrappy" vegetation found bordering fields or roads has been cleared.

Heaping climate change into this mix is another, gargantuan, pressure on insects. It can be hard to tease apart the causes of decline, but Forister and other entomologists are sure that global heating is already sinking its teeth into the insect world. "It's habitat loss, toxicity in remaining habitat and climate change," he says. "It's hard to say which of those is really the biggest thing but it's like a firing range. All those bullets are flying."

Insects are under fire from the poles to the tropics, and there's not much cover to duck behind. The Arctic bumblebee, or *Bombus polaris*, is found in the northern extremities of Alaska, Canada, Scandinavia, and Russia. It is able to survive near-freezing temperatures due to dense hair that traps heat and its ability to use conical flowers, like the Arctic poppy, to magnify the sun's rays to warm itself up. Rocketing temperatures in the Arctic, however, mean the bee is likely to become extinct by 2050. Species of alpine butterflies, dependent on just one or two high-altitude plants, are also facing severe declines as their environment transforms around them.

Farther south, in England, glowworm numbers have collapsed by three-quarters since 2001, research has found, with the climate crisis considered the primary culprit. The larvae of the insects feed on snails that thrive in damp conditions, but a string of hot and dry summers has left the glowworms critically short of prey.

Meanwhile, a forty-year quest from 1969 onward to trap and study the aquatic insects of the Breitenbach, a headwater stream located in the hills of the state of Hesse in central Germany, found a spectacular 80 percent drop in the abundance of mayflies, stone flies, and caddis flies. The waters of the stream have heated up by an average of 1.8°C (3.2°F) in this time, the researchers noted, pointing to a major environmental shift that has had disastrous consequences for the insects.

These sort of losses in Europe have challenged previous assumptions that insects in temperate climates would be able to cope with a few degrees of extra heat, unlike the mass of species crowded at the world's tropics that are already at the upper limits of their temperature tolerance. A team of researchers from Sweden and Spain have pointed out that this shibboleth overlooks the fact that the vast majority of insects in temperate zones are inactive during cold periods. When just the warmer, active, months of insects' lives were considered by the scientists, they found that species in temperate areas are also starting to bump into the ceiling of livable temperature. "Insects in temperate zones might be as threatened by climate change as those in the tropics," Frank Johansson, an academic at Sweden's Uppsala University, glumly puts it.

Bumblebees, those large, furry insects permanently sewn into their winter coats, are at the pointy end of this rising heat. A study by the University of Ottawa in 2020 found that bumblebee populations in North America have nearly halved, with those across Europe declining by 17 percent, in recent decades, with the bees suffering worst in areas that have heated up the most rapidly. Civilization faces a stark future "with many less bumblebees and much less diversity, both in the outdoors and on our plates," says Peter Soroye, coauthor of the paper, who warns that on current trends "many of these species could vanish forever within a few decades."

Some scientists have warned that the correlation shown in this research has yet to prove causation, but there is a broad acceptance that changes in temperature and rainfall could overwhelm insects already facing a barrage of threats. In 2019, for example, scientists revealed the happy news that nine new bee species had been discovered in the South Pacific island of Fiji, only to then immediately note that many of them face climate-related extinction due to their warming mountaintop habitats. "In the future, climate change is going to be the nail in the coffin for quite a lot of creatures which are already in much reduced numbers," says Dave Goulson, the University of Sussex ecologist. "They'll simply be unable to cope with a 2°C rise in temperature and all the extreme weather events that are likely to go with that."

The extra warmth around us not only makes bumblebees swelter, but is also melting the Earth's vast ice sheets and thermally expanding the ocean, causing the seas to rise. Insects are being pummeled by this aquatic advance, just as coastal cities are.

In the United States, the Bethany Beach firefly is unusual as it flashes green, not yellow, with the female of the species acquiring toxins for its self-defense by luring males of other firefly species and devouring them. The firefly is only found in coastal Delaware, with its habitat set to be gobbled up by sea level rise by the end of the century. Farther south, in Florida, the Miami blue butterfly is facing a similar fate, with the encroaching tides clawing away at its favored home of vegetation that fringes sandy beaches.

The talismanic Joshua trees, found only in the Mojave Desert of the southwestern United States, are wholly reliant upon an unglamorous gray insect, the yucca moth, for its pollination, a relationship described by Charles Darwin as the "most wonderful case of fertilisation." The Joshua trees, with their crooked limbs and clubbed tufts, are a regular scenic backdrop in movies and TV shows, but rising temperatures and prolonged droughts are narrowing their habitat. They may all be gone by the end of the century. This would spell disaster for the yucca moth—which purposely pollinates the Joshua tree to provide seeds, a rare food source in this arid landscape, for its caterpillars—but also the other lizards, birds, and insects that rely on the trees for habitat.

Even the Amazon rainforest, that humming trove of insect life, is seeing complex relationships torn asunder. Periodic climatic events called El Niños are getting hotter and drier, which, coupled with human interventions such as deforestation, are spurring more intense drought and wildfires. Researchers were shocked to find this changing regime is causing a population collapse among dung beetles, which are key distributors of nutrients and seeds and important indicator species of the health of an ecosystem. Counts of beetles prior and after an El Niño event in 2016 found that insect numbers had been cut by more than half within the studied forests. The climate crisis is making the Amazon drier, more brittle, and more prone to fires, while also stripping away the unheralded dung beetles that help regenerate burned forests. "I thought the beetles would be more resilient to drought than they were," says Filip França, the Brazilian scientist who led the research. "If climate change continues we'll not only see less biodiverse forests but also make them less able to recover after further disturbances."

Insects are so interlaced with the environment that they acutely feel any jolt to the regular rhythms of life. The disruption to weather patterns, habitat, and even the timing of the seasons is causing insects the sort of confusion that fireflies suffer from blazing artificial light. Spring is being pushed earlier and earlier in the year, unsettling the established life cycle of insects. In the United Kingdom, moths and butterflies are emerging from their cocoons up to six days earlier per decade on average, while in parts of the United States, springtime conditions that trigger insect activity occur as much as twenty days earlier than they did seventy years ago. Most plant and animal species rely on the buildup of heat in spring to set in motion flowering, breeding, and hatching of insect eggs. The reshuffling of the season's dawn risks throwing delicately poised interactions off-kilter, such as birds migrating early only to find a food source isn't quite ready for them yet.

Insects are central characters in this profound change. British scientists who looked at half a century of UK data found that aphids are now emerging a month earlier than they once did, due to rising temperatures, while birds are laying eggs a week earlier. The aphids aren't necessarily growing in number, despite their elongated season, but their earlier appearances means they are targeting plants that are younger and more vulnerable.

This phenomenon is even happening in shaded woodland areas that should act as cool refuges. "If you wanted to go sunbathing, you'd probably go to the beach, or grassland," says James Bell, the ecologist who led the study. "You certainly wouldn't go to woodlands. So we were really surprised to see that climate response there." There is, apparently, nowhere to hide.

The leafing date of trees triggers the emergence of caterpillars, which in turn determines when birds that feed on the caterpillars will lay their first eggs; even a slight shift in this equation can have cascading consequences. Bees typically take their cue to emerge from the temperature, while many plants take their cue to flower from the length of the day, causing another mismatch as springs and winters get hotter and hotter. "There's good evidence here in the UK that under climate change things are warming up early, so we've got all these bees coming out early but not the flowers because obviously the day length isn't changing," says Simon Potts, the University of Reading bee expert. "We're getting this decoupling between pollinators and the plants and that's starting to mess up all these very delicate, very sophisticated food webs."

For some insects, a warmer Britain is a welcome development. In recent years, insects such as the violet carpenter bee and the camel cricket have crossed the English Channel and established themselves, while some native butterflies, such as the marbled white, are hauling themselves out of population declines with a climate-assisted march northward to cooler climes. Flowers such as wild orchids are heading north, too.

In the spring of 2020, Richard Fox, associate director at the Butterfly Conservation charity, excitedly shared a picture on Twitter of a comma butterfly, once mainly found in southern England, spotted near Scotland's Dunnet Head, the most northerly point in mainland Britain. "Butterfly species may not be able to live in hotter places near the Mediterranean in the future but Britain is still a cool, damp place so there's space to colonize for some species," Fox says. Some British butterflies that are less flexible and restricted to a single reproductive cycle a year are likely to fare less well, however. Butterflies are the quintessential "Goldilocks" creatures, requiring surroundings that are neither a little too cold nor a little too hot. "Don't write a butterfly book—climate change will make it instantly out-of-date," Matthew Oates, author of *His Imperial Majesty: A Natural History of the Purple Emperor*, told *The Guardian*.

That said, some adaptation is possible from butterflies and other insects. Researchers have even found that monarch butterflies are developing bigger wings to cope with longer migrations to find a suitable climate for feeding and reproduction. It's also been discovered that if famished bumblebee queens emerge from hibernation a little too early for flowers in bloom, they will nibble holes in leaves, spurring plants to blossom weeks ahead of schedule. But these techniques will mean little when climate change utterly warps the properties of the plants themselves, diminishing them as a food source wherever insects can find them. Rising carbon dioxide in the atmosphere, put there through our prodigious burning of coal, gas, and oil, has been optimistically referred to by some as "plant food," given that it is absorbed by vegetation and helps fuel its growth.

This new diet is for plants much like the menu served up by monocultural farming to insects—a lot of the same thing, not necessarily healthy. Carbon dioxide (CO_2) helps plants grow in the same sort of way a diet solely made up of chocolate cake would aid the growth of a child. Scientists have found that CO_2 can reduce the nutritional value of plants, providing insects with a meal of empty calories lacking elements such as zinc and sodium. A study site in the prairies of Kansas found that grasshopper numbers there are dropping by around 2 percent a year, and researchers felt confident enough to rule out pesticide use or habitat loss as the likely cause. Instead, they concluded that the grasshoppers were suffering starvation via climate change.

Samples of the tallgrass that grows on the study site were collected and stored each year, enabling researchers at the University of Oklahoma to learn that while the overall weight of the grasses had doubled over the past thirty years, likely due to the accumulation of CO_2, the plants' nitrogen content had declined by 42 percent, phosphorus by more than half, and sodium almost completely. For grasshoppers, other insects, and herbivores in general, this is making their meals "more like iceberg lettuce than kale," according to Ellen Welti, the lead researcher. "Increased CO_2 is making plants less nutritious per bite and insects are paying the price."

Not only is climate change potentially causing insects to be malnourished; it also appears to be altering the scent of plants. Pollinators searching for food will note the color and number of flowers as well as the plant's scent, with bees able to recall a fragrance and associate it with certain plants and their nectar content. Scientists who measured the fragrance molecules emitted by rosemary in shrubland near Marseilles, in France, discovered that a different scent was given off by plants that were stressed, which deterred domesticated bees. As climate change stresses more plants by subjecting them to drought and soaring heat, insects may find them not only a bland meal but also unappealing to even approach.

This alteration in plants may be, for insects at least, the most far-reaching symptom of climate change. "There's a lot we don't understand yet but my gut feeling is that plants are being seriously affected," says Matt Forister. "Insects are pretty sensitive to small changes in plants. You have an event like a megadrought and a lot of things will suffer at once."

Not all insects are doomed in a warming world, however. As with all realignments, there are winners and losers, and our attention is more easily captured by thoughts of hordes of marauding insects unshackled by global heating than by a handful of scientists fretting about a declining stone fly on a mountain.

Biblical themes of pestilence echoed beyond the coronavirus pandemic in 2020, with East Africa inundated with its worst plague of locusts in decades. The Horn of Africa had been pounded by rainfall, up to 400 percent above average levels, for the latter months of 2019, aiding the reproduction of locusts. Increased heat is also thought to boost locust numbers, with both factors heavily influenced by climate change. Farmers in Kenya, futilely banging pots and pans to scare off the billions of insects, watched helplessly as the sky darkened with locusts that descended to decimate their corn and sorghum. Separate, massive swarms then broke out in western and central India, chewing up land at a rate not seen in a generation.

A hotter world is likely to bring an array of insect pests and pathogens to attack potatoes, soybeans, wheat, and other crops. A group of American researchers have calculated that yields of the three most important grain crops—wheat, rice, and corn—lost to insects will increase by as much as 25 percent per degree Celsius of warming, with countries in temperate areas hit the hardest. Crop pests also tend to thrive in simplified environments that have been stripped of their predators—another legacy of monocultural farming practices.

In the American suburbs, we will see more emerald ash borers, the brilliantly green beetles native to Asia that were introduced to the United States after a few of them clung to some wooden packaging that made its way to Detroit. The rapacious beetles have killed off hundreds of millions of ash trees across North America and are now establishing themselves in eastern Europe. Milder winters mean the pests will be able to spread farther north, causing further devastation.

Even the domestic environment will see a new influx of unwanted insects, with populations of houseflies more than doubling by 2080, according to one estimate, due to changes in temperature, humidity, and rainfall. But while houseflies can cause illness through the transfer of waste onto food, at least they aren't major vectors of deadly conditions.

It is worrisome, therefore, that there's an expansion underway of mosquitoes that carry diseases such as dengue, chikungunya, and Zika virus. The species *Aedes aegypti* and *Aedes albopictus* are two of the most potent carriers of disease, and they are forecast to spread from the tropics as the climate becomes more agreeably warm for them. Their expanded range could envelop an extra 1 billion people, bringing the blight of deadly mosquito-borne disease to parts of North America and northern Europe that have never had to consider this threat before. "There's not going to be a 'somebody else's problem' 20 to 30 years from now," Colin Carlson, a biologist from Georgetown University who studied this spread, told PBS.

Freezing temperatures tend to kill mosquito eggs. This means that a heated-up planet is allowing the insects to conquer new territories, helping trigger outbreaks of dengue in France and Croatia, chikungunya in Italy, and malaria in Greece in the past decade. These incursions are likely to be vanguards; the Mediterranean region is already a partly tropical region, and as heat and moisture continue to build, the central swath of Europe and even the southern regions of the United Kingdom will be within striking range of a fearsome cadre of newcomers. "If it gets warmer we could get West Nile. Malaria could come back too," says Simon Leather, a British entomologist. "We could see a real change in terms of human health problems."

Our reaction to this threat must be calibrated wisely. Mosquitoes are clearly, by the number of people killed, the most deadly animal on Earth to humans; but in our eagerness to vanquish them, we often deploy weapons with high levels of collateral damage. The chemical compound DDT was developed for widespread anti-mosquito use—before mosquitoes developed resistance and the chemical's pernicious impact on other wildlife led to its ban. A more recent replacement, an organophosphate called naled, is now sprayed on mosquito habitat despite evidence that it is toxic to bees, fish, and other creatures.

As mosquitoes' range grows, it can only be hoped that lessons are learned from places such as Florida, where Native Americans once created shrouds of smoke and buried themselves in the sand to dodge thickets of mosquitoes. Early white settlers resorted to slathering themselves with bear fat or burning oiled rags to keep the biters away, but it wasn't enough. The state was roiled by dengue and yellow fever. "Hordes of mosquitoes suffocated cattle and drove humans to suicide," Gordon Patterson wrote in his book *The Mosquito Wars*. As the space age arrived and the Kennedy Space Center rose from the mosquito-infested marshland of Florida's east coast, the insects refused to buckle to the technological might of NASA. One mosquito even managed to hitch a ride on the space shuttle *Endeavour*, darting around the perplexed astronauts as they entered orbit before being sucked into the filter of a fan and squashed.

But if our fears of a seething invasion of heat-loving insects were to be embodied by one animal, it would probably be the Asian giant hornet. You might have heard it referred to as a "murder" hornet. The bulky, thumb-sized hornet has the demeanor of a cartoonish supervillain, with its tiger-striped abdomen, large burnt orange-colored face, teardrop eyes like a demonic Spider-Man, and a pair of vicious mandibles. Despite a flurry of public concern to the contrary, murder hornets do not murder people; they kill honeybees. The hornets loiter outside bee hives and gruesomely decapitate emerging worker bees, dismembering the unfortunate victims and feeding the body parts to their larvae.

This carnage can go on until a hive is completely annihilated, the crime scene marked by thousands of scattered corpses. In some places, bees do fight back. Bees in the hornets' native range have evolved a defensive tactic whereby a mob of bees will hurl themselves at a hornet that enters the hive, covering the invader in a ball-like mass and then vibrating their flight muscles to generate so much heat, up to 47°C (117°F), that the hornet is roasted alive. Honeybees in Europe and North America, however, are unused to the hornet and are essentially helpless in face of the slaughter.

As its name suggests, the Asian giant hornet (*Vespa mandarinia*) is native to the forests and mountain foothills of East and Southeast Asia. It is commonly mixed up with its cousin, the Asian hornet (*Vespa velutina*), which has found its way to Europe and dismembered so many honeybees in the United Kingdom and France that beekeepers have fretted over the viability of colonies already under stress from *Varroa* mites and pesticides. *Vespa mandarinia*, meanwhile, has launched an assault on the western coast of North America, most likely hitching a ride over on cargo shipping.

Three confirmed specimens were discovered by surprised Canadian authorities on Vancouver Island in August 2019, then another hornet was found farther south, close to the US border. By December, the species was spotted again, this time in the United States, around 19 kilometers (a dozen miles) farther south in the state of Washington. One beekeeper, stung a few times by irate hornets, set the entire colony on fire to destroy it. Another fresh hornet queen, found 25 kilometers (15 miles) southwest of the next nearest find, suggested either a repeated influx from overseas or a vigorous dispersal by the hornets.

By May 2020, with the hornet appearing to have gained a decent foothold on the West Coast, the situation had attracted the attention of the *New York Times*, which ran a story headlined "'Murder Hornets' in the U.S.: The Rush to Stop the Asian Giant Hornet."

For weary Americans in the midst of a harrowing coronavirus pandemic that had paralyzed normal life and precipitated mass unemployment, the prospect of murder hornets—a nickname used by some in Japan for the creatures—advancing across the country was confirmation that 2020 was cursed. "Murder hornets. Sure thing, 2020. Give us everything . . . we can take it," tweeted Patton Oswald, the comedian and actor.

Panicked citizens started to butcher anything that vaguely resembled an aggressive hornet, including wasps such as the cicada killer and the great golden digger wasp. Even bumblebee queens were victims of botched amateur pest control. Entomologists' email inboxes started to fill up with images of suspected Asian giant hornets, invariably misidentified. "My colleagues in Japan, China and Korea are just rolling their eyes in disbelief at what kind of snowflakes we are," Doug Yanega, a leading entomologist in California, told the *Los Angeles Times*. The mistaken killings echoed the experience of the United Kingdom, where isolated sightings of Asian hornets in Devon and Cornwall prompted the destruction of European hornet nests by overzealous homeowners. Similar, too, has been the unsolicited advice given to wildlife officials on how to deal with the invaders.

Chris Looney, an entomologist at the Washington State Department of Agriculture, says he received enthused offers to help destroy the hornets from people hundreds of miles away, as well as several "completely nuts" suggestions on how to dispatch them, including a memorable idea to put a volunteer into a protective suit, slather it with sticky material, allow hornets to land on it, and then douse them with toxic pesticides. "I'm not sure if they were taking the piss or serious but that was definitely the most hair-brained idea we've had," Looney says.

Looney, who responded to the threat by venturing to spots in the border city of Blaine, Washington, with makeshift hornet traps of jugs filled with orange juice mixed with rice wine, is less worried about mistaken identity than the potential spread of the hornets across the United States. Some entomologists have been dismissive of an eastward march, pointing to the inhospitable cold winters of the prairies and the towering barrier of the Rocky Mountains, but Looney isn't as sanguine. In an initial analysis of potential dispersal of the hornets, Looney and some colleagues found that the insects could spread down the coast to California's Bay Area, while also reaching as far north as Anchorage, Alaska.

The middle section of the United States lacks suitable murder hornet habitat, but the East Coast would be prime territory for the bee killers. All it would take is for a hornet queen to burrow into some potting mix or other cargo that is loaded onto a train that then trundles cross-country to New York. "This could happen," says Looney. "It's worrisome. We don't know how well they will establish themselves, but if you have a 300-hive apiary of honeybees you could face losing some of them, or maybe all of them. We just don't know yet." The hornets will likely be aided in their colonization by climate change, which is gradually making habitat more conducive for them. And because they nest in the ground, the hornets will also be able to better escape extreme heat than honeybees, which face being exposed to heat waves that could sap their ability to fight off infections and forage for food.

Climate change could help turbocharge the pace of the hornet's advance, similar to the astonishing travels of the Asian hornet in France, where it has moved at nearly 80 kilometers (50 miles) a year since arriving in the early 2000s and is now found in the Alps. Not much is known yet about how far Asian giant hornets can disperse—the spaced-out discoveries in British Columbia and Washington suggest separate introduced lineages rather than the offshoots of the same colony—but a range expansion similar to *Vespa velutina* would spell trouble. *Vespa velutina* queens attached to a flight mill, a laboratory device that essentially functions as a treadmill for flying insects, have demonstrated their ability to fly for 200 kilometers (124 miles) before giving up. "It stretches credulity that they would do that distance even over a week," says Looney. "But it does mean that they can fly quite a bit if they don't find something to their liking." Still, scientists don't know much about what drives the dispersal of the hornets or if the United States will soon be inundated with them to the extent western Europe increasingly is with the Asian hornet.

While this invasion won't be mounted against humans, an increase in giant hornets will invariably mean an increase in extremely painful stings. Get stung a few dozen times and there's a decent chance you will die, a fate suffered by dozens of people in Japan and China. In 2013, at least twenty-eight people perished in Shaanxi, the northwestern Chinese province where the Silk Road originated, from a spate of stings that experts say are becoming more common.

Conrad Berube can vividly recall the searing pain of the giant hornet's sting after gaining the distinction of being the second person in North America to be stung by the creature. His colleague had conducted initial reconnaissance on the nest that had established itself in Nanaimo, a city perched on the eastern extremes of Vancouver Island, and retreated after being stung. Berube was then drafted to deal with the new menace, choosing to approach at nighttime when the hornets would be less active. The beekeeper and entomologist added two extra layers of clothing underneath his normal beekeeping overalls, Kevlar joint guards to his wrists and ankles, and wriggled into body armor normally used, he says, "for things like chainsaw protection or zombie apocalypses."

With trepidation, Berube approached the nest, which was wedged in the dirt of parkland amid a residential area. He was then set upon by several hornets and stung four times across the top of his thighs where the fabric had stretched tightly across his legs. He subsequently found lanced stingers embedded in his leather gloves.

Unlike honeybees, an Asian giant hornet can sting repeated times, delivering perhaps ten times the venom of their favored prey, comfortably enough toxin to kill off a dozen mice. "The pain was like having red hot thumbtacks driven into my flesh," Berube recalls.

Nasty pus-like welts formed around the hornets' stings and Berube suffered muscle aches that made walking up and down stairs painful.

Take on enough venom and renal failure and death can follow, although Berube is philosophical about his injuries. "It's important to keep in mind that it's a defensive mechanism," he says. "I was the invader and they were just defending their brood."

The sting of the Asian giant hornet does not feature on the Schmidt pain index, a scale of hymenopteran stings developed from the excruciating personal experience of Justin Schmidt, a veteran entomologist at the University of Arizona who has been stung by everything from paper wasps—"like a single drop of superheated frying oil landing on your arm"—to bullet ants—"kind of the ultimate, you're writhing in pain for 12 hours." Schmidt says he's never been stung by a hornet, giant or otherwise, but says going by the anguished descriptions of colleagues, it ranks at around a three—the scale goes from one to four. "Three is definitely miserable," says Schmidt, who first developed the index in the early 1980s. "It depends, of course on where you are stung. If it's the eyelid or nose, it's clearly going to be a lot worse."

In Nanaimo, stifling a stream of shrieked expletives, a stung Berube grabbed a carbon dioxide fire extinguisher, returned to the nest and blasted his hornet foes, stunning them. He then picked up the hornets and dispatched them by tossing them into an alcohol preservative. More workers emerged to attack but were also doused in CO_2, allowing the team to dig out the nest and break apart the combs containing the young. In all, around 150 to 200 hornets were vanquished.

Berube is unperturbed by the alarm provoked by the Asian giant hornets, preferring to tell worried beekeepers to "keep calm and keep bees." Another, less tranquil aphorism is found on a poster distributed by the Entomological Society of British Columbia, which has a 1950s alien invasion–style illustration of hornets larger than helicopters tearing apart skyscrapers as people flee in terror. The poster advises people to "Slap!" a hornet with a stick to kill it, "Snap!" a photo of it, and "Zap!" an email to the society with the image. During a hornet encounter, the poster advises, "If stings have been nil it's best to stay still, but once you have been stung, cage your eyes and just run."

It's natural to get squeamish over the idea of a squadron of murderous hornets or the idea that those ever-durable cockroaches will march on despite the surging heat. The genuinely scary part of all this, though, is climate change itself, an existential threat we have brought upon ourselves and all other living creatures that we still, despite decades of increasingly frantic warnings, move too sluggishly to avert.

A little while ago, Berube was listening to a podcast where the question was raised, "Well, what murdered the murder hornet? That's what we should be scared about." It's significant to Berube that the answer, in part, is the CO_2 that he blasted the hornets with. "Carbon dioxide is one of the drivers of climate change and that should be much more on the minds of people than murder hornets," he says. "They should be making changes in all of their activities that contribute to climate change rather than worrying about murder hornets."

But as we've reacted so grudgingly and ponderously to the menace of flooding, storms, and droughts that can spark civil unrest and even wars, what hope is there that the plight of insects will spur us on? A more realistic goal is a concerted effort to restore complex, connected insect-friendly habitat and ensure that it remains largely toxin free, in the hope that this will at least parcel out a little time and space from the onslaught of the climate crisis. We don't have much time left to play with. Although climate change can often feel like a drawn-out, almost imperceptible rearrangement that far-off generations will have to deal with, it is also punctuated with lacerating reminders that it's already well underway.

*

AUSTRALIA IS A COUNTRY hardened to extremes, its environmental life cycles shaped by millenia of fire and regeneration, of baking sun and torrential downpour. Any suspicion it is moving beyond previous climatic bounds is invariably met with a well-worn line from Dorothea Mackellar's 1908 poem *My Country*, which reads: "I love a sunburnt country, A land of sweeping plains, Of ragged mountain ranges, Of droughts and flooding rains."

Still, something did immediately seem a little untoward as Australia moved into summer 2019, with the odd sight around the national parliament in Canberra of honeybees staggering around, as if intoxicated, as others lay dying. Cormac Farrell, the Australian parliament's head beekeeper, had to explain that the bees were drunk on nectar that had fermented due to the extreme heat. Sober bees inside the colony barred entry to the drunks, leaving them to loll around or die from alcohol poisoning.

From this idiosyncratic start, the summer took a boiling and deadly turn. The year 2019 was Australia's hottest on record, breaking the record set six years earlier. In 2013, Australia's Bureau of Meteorology had to come up with a new color for its weather forecast map—a shade of incandescent purple—because temperatures were expected to reach an unprecedented 52°C (125.6°F). The top five hottest years on record here have all occurred since 2005, with a prolonged drought well entrenched as the roasting summer of 2019 began.

The bushfire season started early, in September, with lightning strikes sparking embers that moved swiftly through the parched vegetation. While the fires, aided by favorable winds, swept the vast country, they were concentrated in the well-populated southeast, igniting areas across New South Wales, Victoria, and South Australia. Sydney was enveloped in smoke, its sparkling harbor and opera house shrouded in the pall, triggering fire alarms and hacking coughs. Air quality in the city briefly became the worst in the world, which necessitated people, in this pre-coronavirus era, to don face masks when venturing outdoors.

Somehow, the scenes became more apocalyptic. As fires roared into coastal towns, terrified residents and vacationers fled to the beaches. A picture taken of the beach in Malua Bay, south of Sydney, captured a remarkable scene of a clutch of people huddled in torment as a lone horse stood in the sand in the center of frame, all bathed in a Hades-like red glow. Further down the coast, in the town of Mallacoota, in Victoria, thousands of trapped people had to be plucked from the beach by the Australian navy.

By January, an area of land larger than Greece had burned, more than 30 people had died directly from the fires, and a further 400 had perished from smoke inhalation. Thousands of homes had been razed. Fires have always been a regular part of Australian life, but this hadn't happened before. It felt different. Previous bushfire disasters have single-day names—Black Saturday, Ash Wednesday—but this rolling catastrophe simply became known as the Black Summer. The fires caused a particular cataclysm for wildlife as the flames tore through trees and grasslands. Large tracts of rainforest, never part of the traditional fire regime, burned up for the first time, scorching the biodiverse inhabitants. Australia is one of the few megadiverse countries on Earth, with around one-tenth of all species in the world occurring on the continent. Millions of years of isolation have sprouted unique branches of life here, from the pugnacious Tasmanian devil to the egg-laying, mammalian platypus. The bushfires torched a rare treasure trove of life.

More than a billion animals perished in the fires, according to ecologist Chris Dickman, at the University of Sydney. "It's a monstrous event in terms of geography and the number of individual animals affected," Dickman acknowledged. The disaster pushed several threatened species to the brink of extinction, including those among Australia's 250,000 insect species, of which only around a third have been named.

Scientists fretted over beetles with small distributions that would be clinging to leaves as they erupted in flame, and of aquatic insects wiped out as waterways clogged up with ash. One striking species, the Australian alpine grasshopper, turns a brilliant turquoise at the right temperature and was left marooned by the fires. It may take years to ascertain whether certain species have been rendered extinct, but entomologists point out that any large loss of insects will stymie the restoration of burned forests due to their key role in dispersing seeds, recycling nutrients, and nourishing the soil.

A unique flavor of this tragedy unfolded on Kangaroo Island, an outpost of rugged natural beauty that lies off the South Australia coast. Mostly free of disease or pollution, the island is renowned for its abundant wildlife, including koalas and its own subspecies of kangaroo. It's also a haven for beekeepers and believed to be the last place in the world where honey from pure-stock Ligurian bees is made. These honeybees, *Apis mellifera ligustica*, are native to the Alps of northern Italy, but made their way to Australia in the late nineteenth century. The last pristine colonies free of crossbreeding, achieved through isolation and local laws, are found on Kangaroo Island. As well as a soft and floral honey for the table, the toil of the Italian bees is also fashioned into beauty and skin care products.

Peter Davis grew up on Kangaroo Island and initially kept bees as a sideline on the family farm, marveling at how they adapted to changing food levels and temperature better than other honeybees. This hobby grew into a major vocation, with Davis's business, Island Beehive, now extracting around 100 metric tons (110 US tons) of Ligurian honey a year, making it one of the largest organic honey producers in Australia.

Bushfire is a way of life on the island, much like the rest of Australia, and Davis wasn't overly alarmed when some flames started licking at the bushy mallee eucalypts that dominate the island shortly before Christmas 2019. But a change in wind conditions helped escalate the situation beyond firefighters' control by January 3, imperiling humans and Ligurian bees alike. "Fires are a fact of nature and we have always been alert to this fact but we did not expect the extent of these fires," Davis admits. He frantically shifted hundreds of hives to safer ground, but the advancing fire was so ferocious it incinerated more than 500 colonies over the course of a day. With so much burned vegetation on the island—around a third of the landmass was charred—the fleeing bees had no safe haven to escape to.

Ever since this trauma, Davis has been directly feeding his remaining colonies, which are no longer producing honey. The sugar gum, *Eucalyptus cladocalyx*, is a main food source for the bees, but it was immolated by the fires, meaning it may take more than a decade for things to fully recover. Beekeepers across the country, wracked by years of drought and fire, have warned of domestic shortages of honey.

The hand of climate change is often overlooked in episodes such as these, particularly when the issue is transplanted from the realm of science to pitched political battle. In a small selection of countries—primarily Australia and the United States—climate change has been treated as a sort of partisan viewpoint rather than a unifying scientific challenge for us to overcome. The tragedy of this denial and obfuscation is that many people, and many species, will perish needlessly because political leaders have wallowed in cowardice, vested interest, and ideological posturing for decades.

Climate scientists have repeatedly concluded that hot, dry conditions create a tinderbox-like scenario for vegetation, desiccating the soil and multiplying the amount of flammable fuel for bushfires. An assessment following the Black Summer found that the high-risk conditions that led to the burning were at least 30 percent more likely than they would be in a world without global heating. But the power of all our ingrained biases is such that we can see the death and destruction predicted by science in front of us, in real time, and still find another explanation. Davis himself blames the strength of the fires on regulations that hinder the ability to remove fire-prone brush and trees that could threaten life and property in a conflagration. Bushfire experts say that creating fuel breaks around homes can help save them from blazes but that this sort of hazard reduction is limited in its effectiveness and that the weather—and climate—has a far greater influence on the severity of fires.

To paraphrase Katherine Hayhoe, a climate scientist at Texas Tech University who does a lot of outreach with skeptics, what we personally believe ultimately doesn't alter scientific reality. A thermometer isn't conservative or liberal or socialist. As temperatures have increased, scientists have found that fire seasons have lengthened significantly across a quarter of the world's vegetated surface, including the western United States, southern Europe, and the Amazon. Combined with natural variations, the warming of the planet will continue to expand the window of opportunity for the homes of people, insects, and other creatures to burn.

Climate change's cruel dexterity means that over decades it can liquefy the glacier homes of stoneflies, over years strip plants of nutrients needed by grasshoppers, or over just an hour or two barbecue a species of rare bee. For all the other damage we are currently inflicting on insects, at some point the insect crisis might be more easily seen as one of the many limbs of the climate crisis.

Peter Davis's bushfire ordeal is partially captured on a shaky video taken by his son. The shot is from the inside of his son's property, looking out of large windows at a crimson scene of trees, cars, and a garden swing engulfed in furious flames. At one point a piece of debris clatters against the window. The house appears hopelessly surrounded by a raging inferno.

"It's going to be massively hot in here in a sec," says Brenton Davis, Peter's son, before yelling a warning to his brother to not go into the bathroom because it won't be possible to escape it. He opens the door a crack to spray water from a hefty-looking hose at the encroaching flames, part of a valiant effort by Davis and his two sons to save their property, as well as their neighbors'. The video resumes the next day, as Brenton Davis tours the damage, including the blackened remains of two smoldering vehicles. Somehow, the house survived. Many of the bees did not. "Yeah, well, we tried," he says, by way of commentary.

The bushfires would continue in various parts of Australia for another two months before finally being doused by a slew of rainfall that was met with widespread jubilation. Within a few months, however, the now familiar images of furious infernos, charred towns, and *Blade Runner*–like orange skies were back, this time in the western United States. An area the size of Connecticut burned, several dozen people died, and smoke plumes reached higher than the standard altitude of commercial aircraft, blotting out the sun from the choked Bay Area to the defrosting Glacier National Park, and further still, as far east as New York. It was the biggest year of fire the western states have ever seen, fueled by global temperatures that have risen, on average, by around 1°C (1.8°F) since the dawn of the industrial era.

More, and worse, is to come. Scientists expect the world to warm up by around 3°C (5.4°F), possibly significantly more, by the end of the century. There is no settled "new normal," rather a ceaseless escalator ride through fresh extremes of heat, fire, flooding, and species extinctions until we finally decide we've had enough. In time, 2020's year of incandescent fire will not feel so freakishly remarkable, after all.

6

The Labor of Honeybees

If you follow the full sweep of Silicon Valley, southeast along the tapering edge of San Francisco Bay, you pass the Frank Gehry–designed Facebook campus at Menlo Park before you get to Mountain View, site of Google's imposing glass headquarters. From there, Apple's circular home, squatting like a giant, futuristic bagel from space, is a short hop away in Cupertino. Go farther still, beyond San Jose and its tangle of suburbs and highways, and you swap the epicenter of one behemoth that shapes hundreds of millions of lives for another.

What Silicon Valley is to the arena of technology and social media, the neighboring Central Valley is to the world of industrialized, spreadsheet-efficient farming. Both of these Californian juggernauts have redefined our current era with a heady combination of ruthlessness and innovation. Stretching more than 724 kilometers (450 miles) from the Cascades in the north to the Tehachapi Mountains in the south, Central Valley runs down the core of California and is one of the most agriculturally productive regions on the planet.

Once the floor of an inland sea, the fertile soils of the valley now provide the United States with 40 percent of its fruits, nuts, and vegetables, pumping out gargantuan quantities of strawberries, grapes, lettuces, tomatoes, and oranges. Without this strip of horizontal land, there would be no raisins, olives, peaches, or figs grown in any meaningful quantities in the United States. This hub's influence reaches well beyond its state or country. Ever since gold rush miners switched to growing wheat in the 1850s, the Central Valley has pioneered now-common farming techniques.

Early visitors were awed by steam tractors, an innovation followed by a procession of new equipment sufficient to fuel a steampunk dream—mechanical cotton pickers, sugar beet harvesters, and tomato pickers. Valley farmers quickly mastered irrigation and then resorted to sucking water directly from the ground, an environmental misadventure that is causing the land to sink at a rate of 5 centimeters (2 inches) a month in places. By using cheap labor, developing new plant varieties, and engulfing the land with an expanding arsenal of pesticides and fertilizers, growers in the Central Valley have fashioned a system worth more than $43 billion in food production and handed the world a blueprint to intensive farming that maximizes output and profits.

This mix of ingenuity and raw power has seemingly enabled farmers to make the land bend to their will. But even this colossus still hinges on a small, buzzing variable that becomes more precarious every year—the honeybee. Industrial farming requires an industrial number of pollinators, with almond growers needing more than anyone else. California produces 80 percent of the world's almonds, and the industry is ambitious for further growth. Almond orchards already blanket 473,000 hectares (1.17 million acres) of the Central Valley, an area that's larger than Delaware and has doubled in size in just twenty years. Another 121,406 hectares (300,000 acres) is set to be turned over to almonds in the coming years.

Almond trees require cross-pollination—the transfer of pollen from one tree variety to another—to produce any nuts, all within a short stanza of time each February as buds, and then snow-white blossoms, emerge. Unhelpfully, this is a time when bees are naturally dormant due to the winter chill, meaning the ones that pollinate the almonds must be roused like slumbering emergency workers who weren't expecting to do the night shift. "We're trying to do something awkward," says Charley Nye, a research beekeeper at the University of California Davis.

Each acre (0.4 hectares) of almonds needs two beehives to achieve the pollination task. This means that almost all the world's almonds materialize from the efforts of 2.34 million beehives, containing around 30 billion bees. The extra planned acres of new almond orchards will raise this demand by another 600,000 hives.

California has only around half a million beehives within its borders, so the shortfall is bridged by a remarkable feat of insect-centric chauffeuring. Every year, around 85 percent of all commercially kept honeybee colonies in the United States are loaded onto trucks, strapped into place, and driven hundreds, sometimes thousands, of miles to the Central Valley. For just a few weeks each winter, in a sort of honeybee jamboree, a patch of America is crammed with cedar boxes thick with bees displaced from their homes and forced to orient to a new environment: regimented rows of a single, almond crop. It is the largest pollination event in the world, a staggering operation to bring the natural world to heel and fall in step with our own rhythms.

"You feel like the cowboys of the last frontier, driving all night, sleeping on the way," says David Hackenberg, a Pennsylvania apiarist who started keeping bees in high school in 1962. He used to haul as many as 2,000 hives to California and back again each year.

This endeavor can sometimes go awry. In 2019, a bee-laden truck returning to Montana from California flipped over, disgorging 130 million bees, which then zipped over the heads of horrified motorists. Firefighters in full protective gear had to tackle the swarm. A year later, thousands of bees attached themselves to a woman's car in almond growing country. When her frantic attempts to shake them off by driving down the highway failed, she ended up at the local fire station, where the bees promptly transferred themselves to the vehicle of a startled fire chief.

This unofficial festival of apiarists is known as the "SuperBowl of beekeeping" due to its scale and the generous pollination fees paid by almond growers. But it's not the only game on the calendar—many of these bees will be back on the truck to rumble across the country to pollinate melons in Florida, apples in Pennsylvania, or blueberries in Maine.

The European honeybee, *Apis mellifera*, is a relative newcomer to the United States but has quickly established itself as an itinerant unpaid contract worker responsible for propping up the nation's food system. Globally, too, there's a growing thirst for the labor of honeybees; according to the United Nations, the amount of agricultural production dependent on pollination has increased 300 percent in the past half century. In Australia, for example, around 1.5 billion bees are sent south to pollinate orchards in the state of Victoria, a process that the country's government classes as the largest movement of livestock in Australian history.

But this epoch of bee dependence is colliding with a morbid reality that honeybees, almost everywhere they are monitored in the world, are being assailed by deadly pests, diseases, and toxic chemicals. Beekeeping itself has morphed from a bucolic pastime based around dollops of golden honey to become a sort of frantic rearguard action to stave off disastrous colony losses while attempting to service an agricultural leviathan's escalating clamor for pollination. "It's not the kind of cheap and cheerful hobby it used to be," says Simon Potts, the University of Reading bee expert.

In China, the native Asian honeybee, *Apis cerana*, has suffered an 80 percent population slump since its distant cousin *Apis mellifera* was introduced in the country in the nineteenth century. The European honeybee was brought to China because it produces more, and sweeter, honey, but it also spreads diseases that have decimated the native incumbent. Several Asiatic bee subspecies are now at risk of extinction, raising fears over the future of native plants that are routinely ignored by the European honeybee. Conversely, *Apis mellifera* itself is also facing challenges in its native range. In 2014, the first comprehensive survey of seventeen European Union members found that countries such as Belgium, Sweden, and Denmark were losing more than 20 percent of their honeybee colonies each winter. The winter of 2017–2018 saw a quarter of colonies perish in Portugal, Northern Ireland, and Italy.

French beekeepers have resorted to feeding syrup to their brood due to unusually late frosts that have dried out flowers, negating them as sources of nectar for famished bees. Colony losses have spiraled from an average of 5 percent a year in the 1990s to 30 percent now, with French beekeepers blaming the warped weather patterns on climate change. Despite a national crackdown on neonicotinoids, there are also lingering concerns over the harm caused by pesticides, even organic ones. "Beekeepers are confused," says Henri Clément, secretary-general for the National Union of French Beekeepers. "They are forced to own more hives to produce less and are forced to make many swarms or buy swarms to replace the dead and maintain the herd. Production has collapsed."

Clément says the situation is an "emergency" that warrants "a real policy in favor of agroecology, with a real drop in the use of pesticides, the return of trees and hedges and more diversified crops with better consideration of pollinators and biodiversity." A study on colony losses on the continent found that bees that foraged solely on orchards, oilseed rape, or maize suffered greater winter losses than colonies with a broader menu of plants. The situation, should it worsen further, threatens to destabilize fundamental aspects of life in Europe. "If we lose the bees, we lose fruits, vegetables, even grains. And without those, we begin to lose birds, mammals, and so on," says Clément. "Bees are a cornerstone of biodiversity."

In the United Kingdom, the long-term trend is severe. Between 1985 and 2008, there was a 54 percent decline in honeybee colonies, according to the University of Reading. There are only enough honeybee hives to pollinate around a third of British crops, meaning much of the burden falls on wild bees, such as ground-nesting bumblebees. "Even if you theoretically could put every honeybee hive in the right place at the right time to pollinate, we'd still be massively short," says Simon Potts, the university's bee expert, who adds that while the total number of honeybee colonies has increased in parts of the world, "we have this massive gap, certainly, in Europe and North America."

Potts has studied bees for the past three decades and has lately been drawn into the political realm to press home the urgency of pollinator loss. He once staged a breakfast with invited British politicians where there was barely any food on offer—no jam, no marmalade, nothing that would require a pollinator. This stark point does register with the general public. Potts finds himself involved in unprovoked debates about bee loss with regulars in his local pub, an indication that bees have become the most prominent and relatable symbol of the insect crisis in the public consciousness.

People may recoil at the thought of a bee sting, but broadly, most of us are vaguely aware of the creatures' importance and that something is amiss in their world. Our concern may even be stoking warmer feelings toward bees. "Twenty years ago if I got a call from someone with bees in their backyard they'd want to know how to kill them," says May Berenbaum, the University of Illinois entomologist. "Now they want to know how to help them."

A key turning point in the way we regard bees came with the emergence of a sinister threat in 2006. Worker bees started vacating their hives en masse, 30,000 or 40,000 at a time, leaving their queen and young behind, instantly crippling the colony. Beekeepers have long dealt with dead bees, their bodies tossed out to the ground when healthy hive members perform an unsentimental cleanup, but here there were no bodies, no evidence of a crime other than one of absence. It seemed we had entered a hideous yet baffling new phase.

Hackenberg, the beekeeper from Pennsylvania, has gained a minor celebrity status within beekeeping circles for being one of the first to experience what was soon christened colony collapse disorder, or CCD. In November 2006, he was inspecting around 400 hives he kept just south of Tampa, in Florida, when an odd feeling of dread descended upon him. "My son was on the forklift truck, I was smoking the bees and there were hardly any bees flying around," he says. "There was an eerie feeling that something was wrong so I started jerking the lids off and there's no one home, no bees in the boxes."

Flabbergasted, Hackenberg dropped to his hands and knees and started crawling around in the gravel, searching for the bodies of dead bees. "There was nothing," he says. "There weren't enough bees from 400 hives to fill a 5-gallon bucket." The beekeeper tried to tell his son what was wrong, but was apparently in such shock he just started stuttering. "I have never been lost for words before in my life," Hackenberg said. That year, he lost 80 percent of his hives to this affliction and soon realized that others were suffering the same abandonment.

Reports of CCD filtered in from Florida and Georgia, with the condition touching twenty-four states in all by the end of 2007. Before long it made a leap overseas, to Switzerland and the United Kingdom. John Chapple, a British beekeeper who first raised the alarm when he lost all of the fourteen colonies in his west London garden, has a more poetic name for the disappearances—"Mary Celeste syndrome," a reference to the ship that was found adrift and inexplicably deserted, despite being perfectly seaworthy, in 1872.

Before long, this mysterious syndrome that was emptying beehives everywhere was regarded by some as the dawn of a bee apocalypse. In 2013, a cover of *Time* magazine featured a solitary honeybee with the heading "A World without Bees." The Whole Foods grocery chain, in an attempt to highlight the importance of honeybees, temporarily removed all food reliant on pollinators from a Rhode Island store. Of their usual 453 items, 237 vanished, including apples, avocados, carrots, citrus fruits, green onions, broccoli, kale, and onions.

What causes CCD is still a point of debate. Scientists have theorized it is triggered by disease or pesticides or stress or poor nutrition or a combination of these things. Hackenberg is adamant that neonicotinoids, the bane of insects in so many ways, are to blame. "The only ones debating this are the chemical companies, the science is quite clear," he says.

In one Harvard University study, healthy honeybees were fed high-fructose corn syrup containing imidacloprid, the neonicotinoid that attacks the central nervous system of insects. Within six months, fifteen out of the sixteen test hives had died off. Separate work by French researchers involved attaching miniature radio frequency tags onto honeybees and then feeding them sucrose containing thiamethoxan, another commonly used neonicotinoid. The bees exposed to the chemical were far less likely to return to their hives after foraging.

Whatever the cause, CCD has faded somewhat as a source of existential bee angst, although, unhappily, an acrid group of other threats have jostled to the forefront. In the vanguard is a foe of bees that has devastated colonies around the world despite measuring barely a millimeter in length (the size of a pencil point) and lacking the ability to see or hear.

Varroa destructor—literally, "destructive mite"—was seemingly put on this planet with the sole task of tormenting bees. The mites are composed of little more than four pairs of legs, ideally suited to gripping onto honeybees, and mouthparts that expertly pierce the exoskeleton of their hosts in order to suck out their innards. Specifically, *Varroa* feeds on hemolymph, a circulatory fluid similar to blood, and will drain an organ found in bees that stores nutrients and filters toxins. "It's less like having a mosquito land on you and drain out your blood, and more like having a mosquito land on you, liquefy your liver, suck that out, and fly away" is the vivid description given by Samuel Ramsey, an entomologist whose research has shed new light on the mite's modus operandi.

These tiny arachnids ride around on bees, feeding on them before turning their attention to their offspring. The hexagonal spaces in a hive's wax honeycomb are used as "brood chambers" where bee larvae develop. A female mite drops into this space to feed on the larvae, hiding behind it and emitting a smell that mimics that of a bee to escape detection.

The mite lays eggs that hatch, develop, and multiply again—a mite population within a hive can double every four weeks over the spring—to create more mites to cling onto the weakened bees. This curse can destroy a colony by damaging the bees' immune system and transmitting viruses. *Varroa* originated in Asia, where the local honeybees over time developed a way to remove them from colonies. But when the mites spread to Europe and the Americas in the 1970s and 1980s, *Apis mellifera* had no defence. Colonies around the world have been plagued by the mites, with only Australia able to prevent them from invading. Despite increasingly desperate efforts, a long-lasting treatment for the mites hasn't been found.

The maladies of parasitic mites, poor nutrition, and toxic chemicals often overlap, sapping the strength of hives to the breaking point. "Honeybees might be able to survive many of these problems if the problems occurred one at a time," a US Department of Agriculture analysis notes. "But when they hit in any of a wide variety of combinations, the result can weaken and overcome the honeybee colony's ability to survive."

Colony losses in the United States are among the most punishing in the world. In the winter of 2018–2019, nearly 40 percent of managed honeybee colonies were lost, according to a survey of beleaguered American beekeepers. This dizzying rate of loss means that around 50 billion bees were wiped out in just a few chilly months. That winter was the worst in the thirteen-year history of the University of Maryland survey and represents a steepening drop from the historic overwinter losses of around 10 percent of colonies.

A sort of never-ending beekeeping triage prevents this sort of decline being terminal. The sheer value of honeybees as engines of agriculture is preventing complete collapse even as they are assailed on all sides. Healthy hives are split in two, with a new queen purchased for the new colony (neatly packed and shipped by mail). Before the winter arrives, beekeepers do everything they can to get their queens to pump out more eggs to offset the losses. Meanwhile, in some countries, such as the United States, bees are shifted from one part of the country to another to replenish stocks.

These human interventions, not available to wild bees, have

blunted a complete nosedive in honeybee numbers. "The honeybee is in no way globally endangered," says Dave Goulson. "They definitely aren't going extinct. They are a domestic animal—their numbers are, more than anything, driven by economics." These economic incentives have driven up the number of beehives globally to around 100 million hives, nearly double the 1961 total, according to United Nations figures.

There is huge variation around the world, though—colony numbers have slumped in North America and Europe while climbing in Asia and South America. There's a beekeeping boom in Uzbekistan, Serbia, and New Zealand, while a sense of crisis is gripping beekeepers in Italy, France, and Egypt. This may suggest a period of cyclical peaks and troughs, but there are entomologists who caution against complacency over the seemingly ubiquitous nature of honeybees. "If we hadn't figured out how to reproduce bees at scale, we would already be looking at extinction. I don't use that word lightly," says Alex Zomchek, the Miami University bee expert.

Zomchek points out that honeybees have been around for close to 200 million years, a time span in which the landmasses of the Earth have shifted, the dinosaurs roared and perished, and humanity branched off from other primates to create the wheel, the printing press, and the iPhone. "Yet in the last thirty years we've taken them to a point that were it not for our ability to outbreed their mortality, they'd be facing an extinction event," he says. "In these last few ticks of the clock, they find themselves in peril. We've wedged them into the food pyramid—fruit and vegetables need pollination, which is done by honeybees. We've created a system that is very efficient but also, in a heartbeat, very fragile."

The efficiency of this system has upended long-held expectations. George Hansen used to think, like many people, that the business of bees was honey. When he started Foothills Honey in western Oregon in the mid-1970s, the proposition was simple—keep some bees, make some honey, and sell it to people to slather across their toast or squeeze onto their pancakes. "I thought I was getting into the honey business," says Hansen, his craggy face and pale blue eyes shaded by the brim of a baseball cap with a picture of a bee emblazoned on it. "We never changed our name but we changed our business."

It's January 2020, and I've traveled to a site just outside Modesto, in the heart of California's Central Valley, to meet with Hansen. We sit in his truck as we look out at some of the 7,000 beehives he brought down from Oregon, with hundreds arranged in neat lines in front of us on the concrete of a fenced holding area. In the 1980s and 1990s, Hansen watched as almond growers upped the price they paid per hive for pollination, from $15 a hive to $20 a hive. It got to $50, and Hansen thought that was that, but it kept going. "It's now $200 a hive—if you can get them," he says. "It's unregulated, it's almost like the Wild West in a certain way."

The Wild West nature of this arrangement can even extend to rustling—not of cattle, but of bees. The spiraling value of beehives, along with the glut of almond growers seeking pollinators, is making it increasingly lucrative for thieves to steal bees from unsuspecting beekeepers. There was an explosion in thefts in 2016, with 1,695 beehives stolen compared with 101 beehives a year prior, according to Rowdy Freeman, a Butte County police officer who is commonly referred to as the "bee theft detective." In 2017, the figure was 1,048 hives.

Police have blamed this spike in thefts on a Ukranian gang that swept through several counties in California, leaving beekeepers bee-less in their wake. Two men were arrested and put on trial, accused of a sophisticated operation where forklift trucks were used under the cover of darkness to swiftly load beehives onto waiting vehicles that spirited them away. A scruffy patch of land near the city of Fresno was used as a "chop shop," police allege, where beehives from multiple owners were split apart to make multiple new hives, the rightful owners' names crudely scratched off the boxes.

These crimes are a symptom of a prodigious boom. If a nut could ever have a heyday, then almonds are having one now, several hundred years into their Californian experience. Almond trees were first brought to California by Franciscan friars from Spain in the 1700s and thrived in the sandy loam and Mediterranean climate. An initial almond orchard was planted along the Bear River near Sacramento in 1843, and the crop has, more recently, proliferated. In the 1960s and 1970s, innovations such as almond shakers, sweepers, and pickup machines replaced the toil of having to physically knock nuts off the trees, catch them in tarps, and haul them away.

The scrubbed modernity of almond farming is now on a par with anything seen in Silicon Valley. Head in any direction out of almost any town in the Central Valley and you will confront endless lines of almond trees, a vista of uniformity across every inch of soil with barely a wild, unprofitable flower to be seen.

Denise Qualls, driving amid this landscape, passes a house with little more than an acre of land, all of it crammed with almond trees. "Everyone here has almonds, everyone," she says. Formerly a banker, Qualls noted the growth of the almond industry and managed to forge one of those unusual jobs that tend to sprout from any boom—in her case a bee broker. Qualls juggles a few dozen beekeepers and almond growers in a sort of matchmaking spreadsheet, shuttling between various farms to ensure that bees get paired with orchards. She rents a house in Modesto each January and February to cash in on the almond pollination bonanza. This intense period provides enough income that she can golf, rather than work, for much of the rest of the year.

"I don't think I've seen my husband in fifteen years on Valentine's Day," says Qualls, who sports a striking all-pink beekeeping outfit with veil and keeps Ziploc bags of almonds in the back of her car. A short drive from Hansen's staging area, Qualls and I are chatting at an almond farm as several hundred beehives are forklifted off the back of a truck following their long journey from Texas. An inspector checking for fire ants—an invasive species in Texas that menace crops and native animals—has caught a suspect that stowed away on the truck and has imprisoned it in a sealed plastic tube for further testing.

The pungent aroma of a nearby cattle lot aside, it's a pleasant scene as the sun sets behind the hives—that is until one of the circling bees crashes into my face, plunges its barbed stinger into my upper lip, and adds itself to the death toll. Bumblebees are fairly docile but can sting as many times as they wish—for honeybees, the single act fatally rips out part of their abdomen and digestive tract. Beekeepers are generally unperturbed by stings, so I try to emit only a few pained yelps.

Qualls's job may be lucrative, but it can feel as if it's built on quicksand. It's not a simple job to be a broker in a scenario where the supplier can't guarantee he or she will be able to fulfill the demands of the buyer. "Everyone is concerned because we do have data showing that we have losses," Qualls says. "There are only so many bees to go around and it's not like you can go to Safeway to buy more of them, right?" She believes that some better remedy for the *Varroa* mite will have to be found to reduce the ratcheting pressure on beekeepers because the current system of pollination is probably not sustainable. "We've got to make changes somewhere," she says. "I don't know where, but we've got to do that. We need a plan."

Bringing millions of bees to California as pollination providers requires keeping them alive and healthy in order to meet the deals struck with almond growers. So Hansen, like other beekeepers, has an array of tricks to keep his insect horde strong. There is precious little for the bees to eat in the desiccated holding area, so Hansen has erected a blue plastic container that looks like an upside-down mushroom. Thousands of bees jam themselves down the chimney of the device and through slats at its base in order to gorge on a protein mix made of brewer's yeast, soy flower, and a sprinkle of sugar and vitamins. The same ingredients are also made into a dough, shaped like a hamburger, and placed within the hives. A separate bee fuel, a carbohydrate made from sugar syrup, is stored in plastic tubs and pumped into the hives via a nozzle, like a motorist refueling their car.

Special attention is paid to the honeybee queens, which carry much of the responsibility for a hive's fortunes. Queens are egg-laying machines, pushing out as many as 2,000 eggs in a single day. Colonies can falter if a queen underperforms, so Hansen keeps a box of queens that acts like a sort of training camp—they are kept in tiny cages and fed by other bees through screens. The extra queens are used if Hansen splits a large hive into two smaller colonies and requires a new leader. They are also used as substitutes as Hansen and his team of ten workers move through the hives to manipulate them into prime pollinating condition. A key target of this work is any queen bee that is laying small, erratic batches of eggs. "If we see that we will pinch her and introduce a new one which is young and vigorous and will take care of the colony," Hansen says. "That will build the colony into something that is economically valuable—if I didn't do that it would probably collapse."

In years past, Hansen never had to use these methods to keep his bees strong and healthy, but then the almond industry has been good to him. Around 80 percent of his income now comes from pollination services, and half of that money is solely from almonds.

The almond industry in general has marketed itself adroitly—there has been a 250 percent increase in almond milk sales in the past five years in the United States alone—and adds $11 billion a year to California's gross domestic product (GDP), supporting 100,000 jobs. In many ways it is a stunning success story, the triumph of a healthy, tasty nut in a society wracked with concern about rising obesity levels and the various environmental disasters associated with meat consumption. But this growth also raises questions about whether it is pushing the limits of what the natural world can bear. Can almonds be made in a way that doesn't add further pressure on honeybees?

Similar to how we have confronted and bested previous quandaries, the agricultural industry sees salvation through technological improvements. The Almond Board of California, which represents growers, has crafted a plan to aid the pollinators its members depend on, including efforts to identify new weapons to quell *Varroa* mites and a plea to landowners to plant more wildflowers amid the featureless rows of trees for bees to forage on and gain a more varied, robust diet. The latter initiative has seen some progress, at least—since 2013, just under 13,800 hectares (34,000 acres) of wildflowers have been planted in almond orchards. This is just a tiny fraction of the space eaten up by almond monoculture, with some growers admitting that they worry about bees spending more time on flowers than their cash crop. But the industry insists it is moving in a progressive direction.

"We take seriously our responsibility to beekeepers and native pollinators," says Josette Lewis, director of agricultural affairs at the Almond Board of California. "It's important to remember, too, that it's not an unnatural thing for almonds to be pollinated by bees." Lewis points out that almond orchards aren't huge users of neonicotinoids and that the industry has cut back on spraying insecticides during the period of blooming. To help aid wild bees, a third of almond farmers are making efforts to plant hedgerows, with half opting for cover crops. Change is coming to the model of sterile monoculture, Lewis insists. "Not all almond farmers subscribe to clean and tidy mode, it's a sign we are evolving our thinking," says Lewis, who adds that farmers' desire to fund research into new methods shows their willingness to embrace sustainable practices. "The vision is sustainable almond orchards as part of a sustainable landscape."

Further innovations may ease some of the frenzy around bee demand. The almond industry has developed a nut variety called Independence that is self-fertilizing, meaning it doesn't necessarily need bees for pollination. A decent breeze is enough to move the sticky pollen the few millimeters to the female part of each blossom and create an almond. The deployment of perhaps half the number of bees currently needed by growers would boost production further.

Moving toward a scenario where crops don't require pollinators wouldn't necessarily be blissful for bees, however. Bees may be currently locked into a cycle of dependency, but at least this creates an incentive for farmers to not allow them to be completely decimated. Crop pollination may be one of the few guardrails preventing a vast increase in the amount of pesticides deployed across landscapes. "Since they are recognized and appreciated by most people, honeybees have helped raise awareness of how farm health is related to our own health," says Ethel Villalobos, a researcher at the University of Hawaii who coauthored a study that found that Independence nuts still require bees to flourish. "We tend to lose sight of the benefits of protecting our natural world. Bees have helped us face that there are choices to be made."

Perhaps the hardest choice surrounds what to do about the plague of *Varroa* mites. The most brutally Darwinian solution would be to let honeybees virtually die out from the mites, with the motley survivors possessing the sort of immunity that would build back the species mite free over hundreds of years. The problem with that plan, of course, is that we would have to eat food in the meantime. And so the toil of beekeepers continues. Like many beekeepers, George Hansen slots a piece of plastic impregnated with a federally approved mitricide into the hive in the hope that bees will shed some of the mites. In 2019, Hansen lost 20 percent of his colonies over the winter, which was his worst-ever toll but still better than the national average. Avoiding complete calamity is a never-ending battle. "In nature, even with hives swarming, there aren't more and more bees each year, there are barely the same amount," Hansen says. "If 50 percent of them fail, that isn't a good enough result for me."

In a second staging area, more of Hansen's blue and white hives are clustered near some barren cherry trees. The hives are being given their final preparations before being placed in the almond orchards. It's a sort of pit stop for bees, with a team of workers, their faces obscured by veils, working on the hives. A blur of bees swirl overhead as the hired workers, all Russian, remove and rearrange the frames. There's a scattering of dead bees on the ground from where the colony has, as Hansen puts it, "cleaned house"—an individual bee has a life span of just a few weeks, requiring a constant regeneration of the hive. A former student of languages, Hansen chats away in Russian to his staff, who cheerily manipulate the hives under the weak January sun.

The work to get these hives into shape began the previous summer with an effort to supercharge the egg laying of queen bees once they became active again. Carbohydrates and proteins are fed into the colonies to build up numbers to the point where the winter diebacks aren't disastrous. Still, there's a sense that only so much is within the power of beekeepers. "We have no foolproof treatments, no guaranteed recipe for success," Hansen says. "All we can do is make the colony as healthy as possible and hope in the long run its natural immunity carries it through. Ultimately, that's all we have."

This new drumbeat of death and artificially spurred replenishment can be unsettling to those of us who romanticize honeybees as avatars of the natural world. There are more than 20,000 species of bee in the world, but honeybees do all the things we expect from these remarkable creatures—they sport yellow and black bands on their abdomen, they do waggle dances, they sting, they make honey. No protest in the era of insect declines would be complete without someone dressed as a honeybee.

Through a more utilitarian lens, however, honeybees are not much different from tiny flying cows or pigs. They are more typically deployed as livestock that we engineer in order to keep our supermarkets stocked with plentiful fruits, vegetables, and honey. We may like to imagine bees within carefree summertime scenes, buzzing around wild meadows, but the reality is that in a growing number of countries, including the United States, there are nowhere near enough wild bees to pollinate certain crops.

The current system may be badly flawed, but without managed honeybees, it would completely fall apart. Hansen recalls the bitter irony of being scolded on his treatment of bees by a friend's teenage child who was munching on an apple—a fruit native to central Asia, introduced to North America by European colonists, and available in stores year-round through a cross-pollination effort undertaken by an army of honeybees, tended to and nurtured by humans.

Yet these concerns aren't mutually exclusive. The reliance on systematic honeybee pollination may be crucial to modern agriculture, bring us a reliable bounty of food, and provide vital income to beekeepers, while at the same time force honeybees to fly a gauntlet of toxic chemicals, debilitating mites, disease, and jarring human interventions. And it's a conflict that beekeepers themselves wrestle with.

David Hackenberg gave up trucking his bees to California a few years ago because he didn't like the impact of Central Valley chemicals on his colonies. "They would come back in such bad shape," he says.

The act of trucking bees around can stress the insects, too, as well as spread diseases from one part of the country to another. Bees naturally move to establish new colony sites, but these swarms travel only a couple of kilometers (a mile or so). It's human transportation that has broken the bounds of distance that bees can reach. "The landscape is now dotted with 'sick' bees and beehives," says Alex Zomchek. "Migratory beekeepers dropping off compromised colonies infect the local hives and local flora." Bees carrying diseases can leave behind residues on plants as they forage that other pollinators then pick up, a process Zomchek likens to the spread of COVID-19. "Different viruses, same mechanics," he says.

For those who push onward, beekeeping can often feel like bailing water out of a giant floating sieve. "We are like heroin addicts," says Jeff Anderson, a beekeeper for the past four decades who runs California Minnesota Honey Farms. "We have to come out here to pollinate almonds, or we are done. Is it good for the bees? No. If you're eating poison 365 days of the year, pretty quickly you're going to get sick."

In many ways, honeybees don't really represent bees at all, with the dividing line neatly summed up by the US government. Honeybees have their own dedicated branch of government for researching and tackling threats, while also within the US Department of Agriculture another, smaller laboratory is responsible for all of the other 4,000 bee species found in the country. If only all threatened insects were deemed commercially valuable livestock, perhaps there would be no crisis in the insect world. There has been a USDA honeybee research laboratory in Beltsville, Maryland, a short drive from Washington, DC, since the 1930s. Housed in a large brick building with opaque windows, the bee lab is on the third floor—the first two floors are dedicated to the study of cows. "Bees are definitely more interesting" is the emphatic opinion of Jay Evans, who leads the research effort at the bee lab. Evans, who has a mop of wavy hair and wears a checked shirt, spends his day fretting about the various things that menace honeybees.

Recent colony losses have been "crazy," according to Evans. "The colonies look pretty good going into winter and then they just don't make it," he says. "It used to be a pretty passive vocation, pretty low stress. Now, there are a lot of stresses."

The bee lab is actually composed of several smaller labs, lined with jars of bees, microscopes, and computers. Bee-related paraphernalia adorns the walls. Beekeepers can send ailing or dead bees to the lab for free to find out what's wrong with them, so there are envelopes and plastic pouches full of groggy-looking honeybees on one bench space. In a fridge, in a sort of bee hotel from hell, bees are kept in transparent vials, each with a *Varroa* mite attached to them. Most of the research here is aimed at better understanding and combating a handful of honeybee disasters, such as *Nosema* and deformed wing virus, which is associated with *Varroa* infestations. Affected bees are only able to grow stubby, ineffective wings, rendering them useless to the colony and quickly ensuring their doom. American foulbrood, meanwhile, is the most offensive to the senses, with infected larvae and pupae smelling a little like dead fish. This bacterial disease kills off bee larvae, is highly contagious, and is often halted in its tracks only by burning all affected hives and equipment.

There's an improved understanding of how these threats are spread, but researchers are mainly scoring wins at the margins, rather than sweeping victories. The enemy is becoming more devilish, too. In 2019, the bee lab discovered that deformed wing virus has become much more genetically diverse, making the development of new treatments far harder. The *Varroa* mite is becoming immune to the various pesticides thrown at it, with scientists toiling to find work-arounds, such as the recent discovery that a bee's gut bacteria can be weaponized against the mites.

The lab has been unable even to stem its own losses—an attractive glass case, framed in wood and placed beside a window, held a colony that mysteriously vanished recently. There have been other oddities, such as queens laying eggs during a warm January, Evans says.

In September 2019, I travel to the lab to see Evans. We don white beekeeping outfits—mine is hopelessly oversized, so I look like a ghost in the midst of drastic weight loss—and stroll to the rear of the bee lab where a clutch of outdoor hives are kept. Evans checks the colonies, puffing smoke from a canister. This technique, designed to distract the hive mind by tricking it into thinking a forest fire is approaching, allows us to inspect the frames covered in bees. All seems well this time, but the threats are never far away. Even at this frontier of honeybee research, the long-term prognosis points to the endless replacement of the mounting bodies.

What we know as European honeybees actually originated in northern Africa. Humans were able to harness their huge numbers, spreading and breeding them across Europe—there are Iberian, Italian-Swiss, and Turkish subspecies—before they were brought over in the early colonial voyages to the Americas in straw skeps. Nonplussed Native Americans called them the "white man's flies." Beekeeping remained something of a curio until a quiet revolution occurred in the Ohio town of Oxford in the 1850s, when a clergyman names Lorenzo Langstroth noticed something about the "Winnie the Pooh trees" that were being cut down and dragged past his property. The honeycomb on these tree-based bee nests were uniform in size—a gap of almost 1 centimeter (0.4 inches) for bees to pass through.

"This was his Eureka moment," says Zomchek, who now lives in Oxford. "That space is the Goldilocks spot, not too big, not too small. Once you have that spacing, you can move those combs to a box and the bees will develop a colony." This one portable invention unlocked the productivity of the land and helped transform America from a country of small family farms to one sliced up among a cadre of agricultural corporations pushing the same sorts of enormous, weedless fields filled with a single-crop, maintained by a barrage of pesticides. "You simply don't have modern farming without that beehive," says Zomchek.

We have since become adept at conjuring up new uses for honeybees. An extreme example of this is a NATO-funded project that has trained honeybees to associate the tempting lure of sugar water with the scent of explosives. Trials in Croatia have shown the bees expertly locating explosives within a 2-kilometer (1.25- mile) radius of their hives, performing even better than sniffer dogs. Not content with using them to prop up our food production, we now expect honeybees to help rid the world of land mines.

This new reality has been both the making and the hammer of honeybees. But all of the focus on a handful of honeybee species has obscured, and even contributed to, the genuine existential crisis in the bee world. The buzzing creatures most at threat of extinction are the bumblebees, solitary bees and other nonmanaged species that make up the vast majority of the 20,000 types of bee on Earth. "Honeybees are a great poster child for pollination as a service but they draw way too much attention," says Philip Donkersley, an entomologist at Lancaster University. "They draw all the conservation attention that people should place on the wild pollinators."

We know less about the status of wild bees than managed honeybees, but the research suggests all is not well. Wild pollinators are in decline in northwest Europe and North America, notes a 2017 IPBES report, which suggests that "long-term international or national monitoring of both pollinators and pollination is urgently required" to reveal the true scale of the crisis. We know, at the very least, that the losses won't be light. The IUCN made its first ever assessment of the nearly 2,000 wild bee species in Europe in 2015, and while admitting its data were incomplete, the world's leading conservation body stated that one in ten species are at risk of extinction. This number contains startling declines—a quarter of bumblebee species on the continent are in danger of being wiped out.

A familiar litany of problems—insecticide use, intensive silage production that razes herb-rich grasslands, the plowing of wildflowers—are menacing wild bees, which, unlike honeybees, do not have dedicated keepers to tend to them nor the belated concern from their tormentors in big agriculture. This blight spans continents. In the United States, there is an anguished effort to prevent the extinction of the Mojave poppy bee, now found in just seven isolated pockets in the Nevada desert. The bee, with black and yellow stripes like a honeybee but with a more slender abdomen, has disappeared as two rare, mutually important desert poppy flowers have declined. On the other side of the country, an analysis of the museum collection at the University of New Hampshire found that fourteen bee species in New England have declined by as much as 90 percent over the past century. The downward spiral of these insects, which include ground-nesting leaf-cutter and mining bees, "raises concerns about compromising the production of key crops and the food supply in general," the researchers warned.

There has been some belated recognition of such declines. In 2017, the US government made its first endangered listing for a native bee species after the rusty patched bumblebee was found to have suffered a catastrophic 95 percent reduction in population. The fuzzy creature, once an important pollinator of wildflowers, cranberries, and apples, has vanished from the grasslands and prairies of the eastern United States and is now found in just a few isolated areas. There are, sadly, many candidates to join this list. In a review of more than 4,000 native bee species in North America and Hawaii, the Center for Biological Diversity found that of those species with sufficient survey data, more than half are declining. Worryingly, a quarter are in danger of toppling over the precipice into extinction.

Other bee species are in greater peril than honeybees. What's more, they perform tasks that their more famous itinerant cousins are unable to do, or are less efficient at doing. The shimmering blue orchard bee, for example, is far better at pollinating cherries and almonds than honeybees. However, this solitary bee, which nests in reeds and abandoned holes to rear its young in chambers constructed of mud, reproduces more slowly and is therefore more expensive for growers than honeybees.

Some jobs are simply beyond honeybees. Pollen, which is essentially plant sperm, is daubed onto pollinators in return for a free drink of nectar, but some plants give this up more easily than others. Tomatoes, peppers, pumpkins, blueberries, and cranberries are all crops that require stimulation, or buzz pollination, to release their load of pollen. Bumblebees are hefty enough to whir their wings to vibrate at around 24,000 beats per minute, creating a distinctive, angry-sounding buzz. This buzzing can produce forces of up to 50G—about five times greater than that experienced by fighter jet pilots. This skill, which honeybees lack, dislodges the pollen and allows for propagation of these foods. A team of bumblebees, their plump furry bodies ideal for attracting and spreading pollen, can therefore prove very useful in a greenhouse filled with tomatoes. Around 2 million bumblebee colonies are exported from Europe each year to greenhouses in more than sixty countries, although this mass transit has backfired in some places.

Chile is the proud home to the world's largest bumblebee. A creature with a shock of fur, it is called the giant golden bumblebee, or *Bombus dahlbomii*, and measures up to 4 centimeters (1.6 inches). Nicknamed the "flying mouse," its numbers are plummeting as it is outcompeted by European newcomers. Public campaigns have been mounted in Chile to save the giant bumblebee, known locally as the *moscardón* and venerated by indigenous people as a vessel for the spirit of the dead. It's a glum irony that the European bees throwing the *moscardón* out of kilter are themselves in danger back in their native ranges. Unlike honeybees, these wild bees cannot be cushioned by sheer numbers; there are at most a few hundred bumblebees in a ground-based nest compared with tens of thousands of honeybees dwelling in a hive.

With such small populations of wild bee colonies, any harmful pesticide, disease, or other calamity that tears through their habitat can spell disaster. It's no wonder that some growers have taken the offbeat step of turning to machine vibrators to buzz-pollinate tomatoes, although it's not a job for the impatient. In one test, it took nearly 12 tedious hours to pollinate 640 tomato plants with a vibrator—that's just more than 1 minute per plant.

Bees perform daily wonders that we benefit from, for no fee. As well as pollinating everything from cashews to grapefruit, bees also provide the beeswax, a sealant they create for their hives, that we humans use in products such as lip balm and the propolis, varnish for stringed instruments, and, in some countries, toothpaste. There is scant evidence that technology is ready to fill the pollination deficit, so the specter of massed human labor as replacement bees is never too far from any imagining of the future of agriculture.

This scenario is already a reality in parts of southwestern China, where rampant use of pesticides and a dearth of natural foraging habitat for bees have winnowed away pollinators from the region's apple and pear orchards in recent years. To replace the lost bees, ranks of farmworkers fan out to pollinate trees in the orchards by hand, using pots of pollen with paintbrushes or sticks with chicken feathers attached to the end. The work to move pollen between blossoms is painstaking but is viable in rural China due to the availability of cheap labor.

It's unlikely that farms around the world will be able to find and afford the labor to do this, however, or that the results will match that of an insect that had coevolved with plants long before the first humans thought about what was for lunch. This raises worrying questions over the pollination of crops in Europe, for example, where most pollination is done by wild bees. Generally speaking, European countries have smaller field sizes than the United States and have retained more bee-friendly hedgerows, meaning they don't need far-ranging honeybees to spew out from strategically placed boxes to pollinate crops.

An ordered arrangement of natural services is unraveling as wild bees die off. Europe will likely start leaning more heavily upon managed honeybees for pollination, which will stave off an immediate food security crisis in the short to medium term but will place it in the sort of bind of dependency that the United States finds itself in. "Increasingly we'll have a reliance just on the honeybee and if we've been taught anything over thousands of years it's that if you start to rely on just one thing to do a job, the second that thing breaks you can't do it anymore," says Donkersley, the Lancaster University entomologist.

There are sixty-eight species of bumblebee across Europe, with around half of them in decline and sixteen on an endangered "red list." Donkersley's personal favorite is the bilberry bumblebee, *Bombus monticola*, a species with a charming orange-red tipped abdomen, only found in a few mountainous areas and upland moorland. In what he calls a "classic entomologist" scene, Donkersley once caught one of these bees in a giant butterfly net as he sprinted across a peak in the Yorkshire Dales. The species' prospects are dwindling, however, due to the seasonal burning of its habitat for grouse shooting.

A systematic pollinator of plants—compared with the random zipping around of honeybees—bumblebees must beat their wings incredibly fast to remain in the air, at around 200 times per second, burning through an enormous amount of energy as they do so. Dave Goulson, the University of Sussex biologist, wrote in a 2010 paper that the animals have the highest metabolic rates recorded in any organism, even surpassing hummingbirds. If a human male devoured a Mars bar, he'd burn the energy off in around an hour; a bumblebee of an equivalent size would use the same energy in just 30 seconds. This voracious need for energy, in the form of nectar-rich flowers, "lies at the crux of the problems facing bumblebees in modern Britain," wrote Goulson, who has forlornly searched southern England for bee species that were commonplace a century ago. With grassland and wildflowers virtually eliminated from the countryside, bumblebees, and wild bees more broadly, are waning badly. A 2019 study concluded that a third of British wild bees and hoverflies are in decline.

The consequences of losing bumblebees would be dire. "Bumblebees are probably the most important natural wild pollinator, right through Europe and across to China and through North America," says Goulson. "There are quite a lot of plants that wouldn't set any seed at all without bumblebees." Still, there is a kaleidoscopic world beyond just the family Apidae—which includes honeybees and bumblebees—that spans bees in an array of colors, most of them unable to perform waggle dances or make honey. Solitary bees, which live alone without queen or colony, make up 98 percent of bee species in the United States, for example, and are globally important pollinators.

Many of these bees would be mistaken for wasps or perhaps unknowable aliens by a good chunk of the public. Megachilidae is a family of bees that have huge heads and powerful mandibles that they use to crunch through leaves, mud, gravel, or wood pulp for their nests. They are found in underground burrows, wood cavities, and even snail shells. Some have perfected living in reeds and are able to cut a leaf to the exact size to plug the nest entrance to prevent their eggs from being eaten. Halictidae, another ground-nesting family, consists of the sweat bees, named for their habit of lapping at a glistening brow for salt. The alkali bee, a distant cousin, is known for its gorgeous iridescent abdominal bands. The minute Quasihesma bee, found in the most northerly tip of Queensland, in Australia, is hairless and can measure a mere 1.8 millimeters (0.07 inches) long. Mining bees can tunnel underground. Vulture bees feast on the rotting meat of dead animals, rather than nectar and pollen, using their sharp jaws to slice into corpses' eyes before stripping all of the body's meat away.

In some cases, these wondrous bees are being indirectly harmed by their more famous cousins. In the summer of 2019, Samantha Alger, a pollination expert at the University of Vermont, and three colleagues came out with a study suggesting that honeybees are transferring diseases to wild bees. The scientists explored more than a dozen sites in Vermont and found that two honeybee afflictions—deformed wing virus and a deadly condition affecting queen bees called black queen cell virus—were higher in bumblebees found less than 300 meters (328 yards) from commercial honeybee hives.

The team also detected the viruses on flowers near the honeybee hives but, tellingly, could not find traces of the diseases on bumblebees or plants farther away from the hives. This finding provided greater clarity to latent fears among scientists that the mass movement of honeybees for pollination is smearing a brew of hideous diseases across wild bee populations. "A huge misconception in the public is that honeybees serve as the iconic image for pollinator conservation," says Alger. "That's ridiculous. It's like making chickens the iconic image of bird conservation."

In 2017, a review of previous studies in this area found mixed evidence, although in most cases honeybees were deemed to have a detrimental impact in terms of spreading diseases as well as outcompeting their wild bee counterparts in the same habitat. A year prior, a scientific paper found that *Varroa destructor* can "nimbly climb" from flowers onto visiting bees, including bumblebees.

These intra-bee headaches are tainting one of the most popular responses to the insect crisis—urban beekeeping. From Detroit to London to Sydney, there has been a boom in the number of people looking to keep hives in their gardens or on rooftops. Enrollments to beekeeping courses have swelled. Scores of businesses have proudly trumpeted the environmental credentials of the buzzy hives placed atop their office blocks. In Berlin, urban beekeeping has become so popular that around thirty people, known as the *Schwarmfänger*, have specialized in dealing with clumps of honeybees found under eaves or clustered on lampposts once budding beekeepers realize they are a little out of their depth. "It's quite hip at the moment, people put up a hive on their balcony somewhere and think they are doing something for nature," Alfred Krajewski, one of the volunteer swarm-catchers, told the *New York Times*.

Anyone can keep a hive, which can explain why beekeepers persist in war-torn Syria or within the depths of refugee camps in Tanzania. This doesn't mean that nurturing honeybees is always a beneficial option. Honeybees, through their sheer numbers, quickly dominate an urban environment that has few available plants, leaving very little food for the outgunned wild bees. People who are drawn to keeping bees are also likely to be the type to plant a helpful variety of flowers for bee forage or to tend to an ailing bumblebee with a teaspoon of sugar syrup. But the act of keeping hives threatens to undo much of that work if it leads to wild bees being robbed of their habitat and infected with new diseases.

Bees are unusual in that they are more diverse in temperate, rather than tropical, regions but this spectrum is undermined if cities are stuffed with honeybee hives. A 2020 report by Kew Gardens estimated that a square kilometer of urban landscape in the United Kingdom could support around seven hives—in parts of London it's now up to more than fifty hives. Poorly managed hives have become "little ecosystems of plagues and contagion," Jane Memmott, an ecologist at Bristol University, told the *Guardian*.

"If you bring in honeybees you may inadvertently starve all the bumblebees and solitary bees," says Donkersley. "There's not much room in cities for pollinators. Bringing in honeybees only adds to that pressure." Even in rural areas, honeybees can cause problems. A group of scientists in Australia referred to honeybees as a "pest" in a recent plea to authorities not to place managed beehives in the country's national parks, citing concerns that the honeybees would spread the *Nosema* disease to native bees and pollinate hated invasive weeds. Once you shift your gaze away from the agricultural realm or from bearded hipsters making rooftop honey, affection for honeybees can drop away steeply.

Honeybees are "apex nectar and pollen gatherers," Zomchek says, that require around 45 kilograms (100 pounds) of honey to get each colony through a year. As it takes 2 million flowers to make half a kilogram (a pound) of honey, the requirements once we start grouping 100 hives, or maybe even 1,000 hives, start to run into the hundreds of billions of flowering plants. "What modern agricultural areas can support this?" Zomchek asks rhetorically. "What is left for other pollinators after honeybees have combed the countryside?"

Conservation work is littered by these sorts of irritations, where minor ecological players such as pandas are lavished with the sort of attention that should rightfully be focused on environmental linchpins such as coral or intrepid engineers like beavers. But this, ultimately, isn't the fault of the panda, just as it isn't the fault of the honeybee. Humans are the ones who have refashioned the planet in awkward and self-destructive ways, while it is our own innate biases that have led to some creatures being prioritized over others.

In the shriveled corner of modern human life that interacts with animals, there are our pets, the occasional backdrop of livestock, and the exotic subjects of wildlife documentaries that are routinely co-opted in advertising and children's books. Honeybees, for better or worse, manage to straddle a couple of these categories—common enough to feel familiar but with the magical twist of somehow making honey. They never asked us to exploit their versatility by spreading them around the world, picking up diseases on the way, to help feed us. Politicians made them props, and lazy marketing made them symbols of conservation. The best we can do going forward is to leverage their profile for the benefit of other pollinators, in the same way "keystone" species such as orangutans and tigers have been used as high-profile totems for their own ecosystems.

In practice, we should perhaps consider the intrinsic value of bees in general rather than just their utility to us. We may not be able to extricate them from the sprawling food production matrix, but we can pause to think about what more we can do to sustain the well-being of these remarkable creatures.

7

A Monarch's Journey

California's year starts with a huge influx of bees and it draws to a close with another mass insect migration, one that doesn't require truck drivers, body-enveloping white suits, or humming hives. Each October, a cresting orange and black wave of monarch butterflies rolls in from as far as Idaho and Utah to mix with the surfers and tech workers on the California coast.

This spectacle draws tourists to groves dotted along the southern half of the state, where the butterflies can be seen thronging eucalyptus trees in the relative warmth of their winter vacation. Around the same time, a separate swarm of monarchs launches an even longer, more gruelling migration, one that spans international borders and multiple generations, journeying from the northeastern United States and Canada down south to around a dozen select hideouts in the Sierra Madre, a mountain range that rises like a rugged spine through central Mexico.

Compared with a modern aircraft firing up its engines and getting you from New York to Mexico City in around 5 hours, a monarch's journey of the same route is a monumental feat. A fragile wisp of a creature, weighing as much as a raisin, is able to complete an odyssey of just under 4,800 kilometers (3,000 miles) using just its wings, air currents, and finely honed instinct. Millions of monarch butterflies somehow pull off this routine every winter, some traveling as far as 400 kilometers (250 miles) in a single day.

Scientists have long puzzled over the navigational abilities that allow monarchs to cover this vast distance. Some answers have begun to emerge in recent years. In one study, several of the butterflies were placed in miniature flight simulators, which helped reveal that the creatures have "light-sensitive magnetosensors" in their antennae that act as a compass to guide the monarchs south to Mexico.

Monarchs flee south to escape the cold and return north, in stages over a few generations, in order to lay eggs on milkweed, their favored herbaceous plant, once it blossoms in spring. Caterpillars emerge to feed on the milkweed, their only food source. Being cold-blooded, monarchs require optimal climatic conditions to thrive, although intricate experiments are necessary to ascertain exactly how the butterflies time their warmth-chasing trips.

Over a twenty-year period, volunteers across North America undertook the painstaking work of placing small, circular tags on more than a million monarch butterflies as they flitted south. Around 13,000 of these markers were retrieved once the monarchs made it to Mexico. Researchers logged the data and realized something intriguing—the monarchs were starting and pacing their migration based on the angle of the sun. Most of the insects took flight when the angle of the noon sun was about 57 degrees above the horizon, no matter where they started their journey. Monarchs quickened their pace in the middle part of the journey, up to 47 kilometers (29 miles) a day on average, before easing back to around 16 kilometers (10 miles) a day as they reached Mexico.

The mysteries of the monarch migration may be hard to pry open, but the fundamental essence of it, seeing them in flight or roosting in trees, is beauty in perhaps its most distilled form: huge, benevolent blankets of butterflies, enlivening the stunted color palette of our quotidian lives with their dashes of orange and black, on an epic quest to gather in lands unseen.

It's an injection of raw nature so huge that it is visible on weather radar, its impact so profound that people living under the aerial pathway turn into active cheerleaders. Minnesota senator Amy Klobuchar recalls how her mother, a teacher, dressed up each year as a monarch butterfly and held a sign reading "Mexico or bust." There are longer migrations in the insect world—a 4-centimeter (1.6-inch) globe skimmer dragonfly was once recorded making an 18,000-kilometer (11,185-mile) journey between India and Africa—but the repeated migrations of the monarchs, in such massed formations, stand alone.

"It's really quite mind bending," says Dara Satterfield, an ecologist who studies monarch migrations. Monarchs heading south tend to congregate into a single amorphous flyway through Texas, something Satterfield witnessed a few years ago in Dallas. "There were hundreds of monarchs just gorging themselves on the frostweed flowers in one garden," she says. "They were so intent on eating and building up fuel that you could pluck them off the flowers like grapes." As the monarchs funnel through Texas, they face myriad perils, from storms and starvation to more direct human-caused calamity. Millions are crushed to death by cars and trucks in certain roadkill hot spots such as Interstate Highway 10, which wraps across the southern United States like a low-slung belt. Even those that complete the journey are vulnerable. With around 95 percent of all monarch butterflies clustered together in a handful of small sites, a single storm or spike in heat can decimate the species. In 2016, a severe storm tore apart thousands of the trees hosting the butterflies, a situation that, combined with freezing temperatures, killed around a third of the monarchs.

As temperatures start to warm up again in spring, the butterflies depart Mexico en masse to find milkweed and mate. Eggs are laid, caterpillars hatch, form chrysalides, and metamorphose into new butterflies, which continue a generational relay north to the United States. By the time the next southward journey begins, monarch butterflies are instinctively returning to trees that even their grandparents never knew. The thought that butterflies attempt such a migration at all is astounding—the fact that generation after generation successfully complete it is almost unfathomable. "It's amazing when they make it," Satterfield says. "We are fascinated by the migration because a difficult journey speaks to us in some ways, on a literary level. It's very human. It has certainly captivated me."

Cruelly, this natural wonder is now so precarious that monarch researchers speak of its demise taking place within just a few decades. A grimly familiar menu of problems—habitat loss, deadly wafts of insecticides, climate change—has chewed through butterfly numbers around the world, with monarchs the unhappy standard bearers of decline.

The mountainous area of central Mexico occupied by overwintering monarchs is traditionally measured in hectares. In the winter of 1996–1997, roosting monarchs covered an area of 18 hectares (44 acres), about the size of eighteen baseball fields. By 2013, this area had shriveled to 0.6 hectares (about 1.5 acres), a size less than London's Trafalgar Square. More than 20 million monarchs can be found in a single hectare. There has been a slight rally in recent years, with 6 hectares (15 acres) of central Mexico hosting butterflies in the winter of 2018–2019, but this apparent recovery promptly dissolved a year later, down to 2.8 hectares (7 acres) of forest.

The parlous state of the western population that heads to California is even more startling—a mighty horde of around 4.5 million butterflies in the 1980s dwindled to just 29,000 individuals in 2019, according to surveys conducted by the Xerces Society. The year prior, a record low of 27,000 was recorded—a population less than 1 percent the size of the historic high. Monarch butterflies are teetering on the precipice, and it won't take much for the diminished survivors to topple over the edge. "If things stay the same, western monarchs probably won't be around as we know them in another 35 years," Cheryl Schultz, a monarch researcher at Washington State University Vancouver, has warned.

Like most insects, populations can sharply fluctuate. In Mexico, numbers have rallied a little in recent years, but the long-term trend points miserably downward. In 2015, the US Fish and Wildlife Service shared the unhappy news that since 1990, close to 1 billion monarch butterflies had vanished, roughly the same number as all the humans in North and South America combined.

The spectacle of millions of tiny orange voyagers draped on trees, collecting in such multitudes that branches buckle and sometimes snap under their weight, may soon become a thing of fading memory and then incomprehension, much like the idea that lions once occupied Europe. "In the next fifteen years we will probably see a very significant decline because their habitat is degrading," says Orley Taylor, a renowned butterfly expert who in 1992 founded a research and conservation group called Monarch Watch. "They represent something unique on this planet and we are in the process of ensuring its obliteration."

The reduction in monarch numbers has been tracked in earnest since the 1990s, with scientists still untangling the reasons for this slump. But the razing of natural habitat for monocultural crop farming and the spraying of chemicals are the most likely suspects. "We don't get as many butterflies down here because Americans spray fucking Roundup everywhere," as Cuauhtémoc Sáenz-Romero, a normally affable Mexican forest scientist, succinctly puts it.

Ominously, Sáenz-Romero has helped identify an even greater long-term threat to the monarchs than even sprays and bulldozers: oyamel firs, the favored trees that the butterflies cling to when they reach Mexico, are at risk of being wiped out due to human-induced climate change. As temperatures continue to rise and periods of drought proliferate, the area suitable for oyamel firs is set to vanish. A 2012 research paper coauthored by Sáenz-Romero found that oyamel-friendly habitat will shrink by 96 percent nationwide by 2090, if temperatures continue their upward trajectory. This shrinkage will increase to 100 percent within the monarch biosphere reserve, completely annihilating the monarch's habitat within their protected zone in the central Mexican mountains.

"There will not be a single square kilometer suitable for these trees," Sáenz-Romero tells me in January 2020, when I visit the monarch reserve. The scientist is fond of wearing berets and multipocketed waistcoats. Twenty years ago, he would routinely need to wear a sweater, too, on chilly days in these mountains. He very rarely needs to these days—central Mexico has been warming up rapidly since the 1970s. "Trees are already dying and it is 2020. If we don't have healthy trees the monarchs will die, for sure. We are on track for the worst case scenario."

Oyamels reach more than 46 meters (150 feet) into the heavens if allowed to grow to a ripe old age, with the needle-covered branches jutting outward as sheltering limbs for monarch butterflies. The trees serve two main functions for the monarchs, first as part of a canopy that acts as a blanket to trap some of the heat released from the ground as temperatures drop, keeping the butterflies cozy. The branches also function as rudimentary umbrellas, ensuring that the monarchs aren't soaked by the rain. Any water pooling on their wings could turn to ice and kill them. It's one of climate change's vicious paradoxes that the heating up of the planet will cause monarch butterflies to freeze to death.

The blight of this disaster is creeping upward in elevation into the oyamel's optimal zone, situated in excess of 3,000 meters (almost 2 miles) above sea level. These firs, like most tree species, are able to adapt to changes in the environment over time, slowly shifting to places with a more agreeable temperature. But the rapid pace of global heating means that the oyamels' preferred climate is zooming up the mountainside well ahead of them, as much as ten times faster than they can keep up with. Most of the trees currently appear healthy at a casual glance, but Sáenz-Romero is quick to jab a finger at the browning extremities of one tree, the drooping needles of another. Lack of water combined with punishing heat weakens the firs, causing them to drain in color, shed their needles, and become overwhelmed by disease.

Some of the trees have telltale patterns of sap on their trunks, a defensive move that indicates that bark beetles have invaded and are devouring the trees from the inside out. Albeit unintentionally, we are recalibrating these forests to expel a butterfly deified in art and verse, in favor of a 3-millimeter-long (0.12-inch) mud-colored beetle that spends most of its time buried within distressed trees, slowly destroying them. "Not all the trees are going to die but they will certainly all be stressed," Sáenz-Romero says. "Monarchs won't be here forever. That statement is very hard to swallow for local people because they live for the tourism dollars. They hate me because I am saying they don't have a future."

The conflagration of a tropical rainforest or the forlorn sight of a polar bear adrift on an iceberg are tragedies we tut over while remaining abstract to our lives. But the winnowing away of the monarchs, a creature familiar to many people's back gardens, feels uncomfortably close. The pang of loss doesn't require validation from any scientist—in Mexico, the crashing of monarch butterflies is easily within the grasp of a person's lived memory. Francisco Ramirez Cruz, well into his 70s when I met him, recalls traversing the alpine forests of his community in central Mexico as a boy, the trees and air thick with butterflies. "We used to see them everywhere but we don't anymore," he says. "We don't see big flocks, just a few here or there. The population is smaller and they arrive a lot later in the year than before."

Ramirez Cruz, known locally by the honorific Don Pancho, served four decades as the elected leader of La Mesa, a small town located in striking, rugged terrain 113 kilometers (70 miles) west of Mexico City. This community runs alongside the Reserva de Biosfera de la Mariposa Monarca, a national park created to protect monarch habitat that was listed as a World Heritage site in 2008.

*

THE RESERVE PLACED boundaries around a sanctuary well known to local people but a mystery to foreigners until relatively recently. American and Canadian researchers spent close to a century seeking the monarch butterfly's Mexican hideaways before finally succeeding.

"Discovered: The Monarch's Mexican Haven" trumpeted a *National Geographic* headline in 1976 after Fred Urquhart, a Canadian zoologist who had spent nearly three decades searching for the elusive spot, finally witnessed oyamels bending under the weight of monarchs in the mountains of Michoacán state. Not far from this site sits La Mesa, which operates as an *ejido*—a sort of collectively owned arrangement where residents share the benefits of the land and its bounty, which in the mountains of central Mexico is mainly potatoes, wheat, and corn.

Traditionally, income has also been derived from logging, with the land shorn of trees right up to the boundary of the butterfly reserve. Illegal logging within the reserve has dropped steeply in recent years as locals started seeing the financial benefits of having an international tourist destination in their midst. But the decline of the monarchs threatens to reverse this equation.

When I met him, Ramirez Cruz sported a neatly clipped moustache and wore a denim jacket topped off with a vaquero hat. Along with his wife, the creator of incredible tortillas, he lives in a tumbledown house with a glorious view of the plunging valley below. The property is supplemented by a small whitewashed chapel, which he painstakingly constructed and which contains numerous depictions of the Virgin Mary. A circumspect man, Ramirez Cruz forged a reputation as a strong advocate for a town where poverty lingers. La Mesa, a place where stray dogs roam and burros still help with transportation, only managed to belatedly get electricity through Ramirez Cruz's persistence.

The trails that wind their way into the butterfly reserve are devoid of people save for the odd *ejido* member paid to maintain wooden cabins intended as accommodation for paying tourists. The cabins have in recent years been completely empty. Visitors have been deterred, locals say, by the lack of butterflies and also the tenuous security situation in Michoacán.

In a gruesome overlapping of these concerns, Homero Gomez, who managed the El Rosario Monarch Butterfly Sanctuary, the largest of eleven community-managed nature reserves at the heart of the protected area, vanished in January 2020. Two weeks later, a man who was feeding cattle on his property spotted a body floating in an agricultural pool. It was Gomez. Gomez gave up his job as an engineer to advocate for the monarchs, which he called "girlfriends of the sun." He posted spellbinding videos of himself shrouded in butterflies on social media. This work forged him as a staunch opponent of illegal logging in the region, despite being a former logger himself. A few days after this grim discovery, Raúl Hernández Romero, a part-time tour guide at the butterfly reserve, was also found murdered.

The true motive for these deaths, or the perpetrators, may never be fully known. Illegal logging has largely been stamped out from the protected area, but its specter lingers. If nothing else, the international outcry over the murders hints at a brutal calculus—if money will no longer flow from the butterflies, it may come from the trees themselves. "If monarchs disappear we will switch to forestry, we will go back to logging," says Ramirez Cruz, adding that other communities are felling monarch-friendly trees close to La Mesa in order to lure tourists to their own butterfly groves. "Other communities are doing logging, so why shouldn't we do that too?"

The monarch's refuge is being crushed by a seemingly unstoppable, circular force. Warming temperatures are killing the trees and imperiling tourism from the monarchs, which may cause the local population to cut down the trees for income, releasing more carbon into the atmosphere, which causes further warming.

Sáenz-Romero has come up with an audacious plan to thwart this death spiral—to literally move the forest up the mountain. By shifting huge numbers of oyamels around 350 meters (less than a quarter of a mile) farther up the mountainside, the surrounding temperature will be more bearable to the trees, allowing them to sustain the roosting monarchs and thus save the great migration. But the plan has its doubters. "Some people think I'm close to crazy," Sáenz-Romero concedes. Still, a motley group of researchers and local landowners have established three experimental planting sites at different altitudes to see how the oyamel would cope with being manually relocated. The seeds for this are obtained from high up on the trees, requiring locals to scale vertiginous heights using little more than a rope, a tight grip, and muttered prayers.

At the highest site, around 3,400 meters (2 miles) above sea level, several oyamels planted four years ago are starting to flourish, aided by the protective shade of nearby shrubs. They have had no adverse reaction from being shoved 400 meters higher in elevation and are healthy compared with those at lower altitudes. The difficulty is that soils make way for rocks as you get closer to the mountain summit, and the area available for trees narrows as the mountain tapers into a peak. The relentless nature of climate change compounds these problems—the average temperature of Mexico, like the rest of the world, is set to surge upward for decades, perhaps centuries. The oyamels are going to run out of mountain here, toppling over the edge into the void as even the summits bake under a blanket of greenhouse gases.

Other enormous forested volcanoes could be suitable for a while—there are nearby peaks that exceed 5,000 meters (3 miles)—but they, too, will eventually be eclipsed. It's as if the monarch butterfly is perched on a ladder where each rung is being systematically set on fire.

A gloomy preview of the monarch's fate is available in the lower elevation oyamel plantings. Ramirez Cruz has several hundred saplings growing in a large wooden tub, covered in netting. He also has receptacles to measure the amount of rainwater, but when Sáenz-Romero picks one up he's astounded to see it is completely empty. The rainy season typically runs from June until October, but farmers need preceding moisture to plant their corn in time for this growing season. There was always a little rain over winter, downpours known locally as *cabañuelas*, but the empty rainwater gauge shows that this year there was zero. Using his forefinger, Ramirez Cruz indicated how the size of the corn he grows is steadily shrinking.

Droughts are extending in Mexico, and rainfall is concentrating into shorter, sharper bursts, a scenario that leads to stunted growing seasons for crops, sicker oyamel trees, and, most disastrously, the sort of landslides that saw several dozen people perish not far from La Mesa in 2010. "This is much drier than I expected, which is concerning," says Sáenz-Romero, as a few of Ramirez Cruz's oblivious turkeys gobble loudly around us. At the oyamel planting at the lowest elevation, at the town of Tlalpujahua, there had also been no nourishing rain in the previous two months. "*Nada, nada, nada,*" Sáenz-Romero mutters, in disbelief. "This makes me think we have even less time than I thought."

Rising temperatures are pummeling monarchs everywhere. Milkweed won't be viable in Texas as the number of days roasting in heat of more than 32°C (90°F) soars. Monarch sites in the Midwest and coastal California are also heating up. "All along the way, if we look at each stage of the population development we can see the writing on the wall," says Orley Taylor. It's a sign of the devotion provoked by monarch butterflies that hundreds of volunteers have aided Monarch Watch, the group founded by Taylor. Many Americans have taken to planting milkweed or breeding monarchs to boost their populations, although once released, these butterflies are less likely to make it to Mexico than wild counterparts that are hardened by the vicissitudes of nature.

This effort has occasionally misfired. A 2015 study found that well-meaning monarch fans had been planting swaths of a tropical variety of milkweed that didn't actually die off during the winter. While the butterflies enjoyed this plant, they had no reason to leave it and migrate southward. Worse, the tropical variety, called *Asclepias curassavica*, hosts a debilitating parasite that weakens the monarchs and shortens their life spans. Most infected butterflies, even if they attempt to migrate, don't make it to Mexico.

This setback is just one kink in the Sisyphean challenge of keeping monarch populations from collapsing. And yet researchers and aficionados won't give up. Taylor fell for monarchs after nearly being killed by its cousin, the sulphur butterfly. He was so allergic to the species that he developed asthma and had to be on medication to reduce the swelling in his lungs. "I would need to sleep outdoors with my back against a tree in order to drain my lungs, it was that bad," he recalls. After a spell working with bees, he started studying monarchs and realized he was fascinated by them, with the added bonus that they didn't pose a threat to his life. "They are an incredible organism that makes us question so much about how life works," he says. "But we are losing ground. We aren't planting the habitat that is being lost each year. It's like what the Red Queen said to Alice—if you want to stay in the same place you have to run as fast as you can. If you want to get anywhere you have to run twice as fast as that. We need to increase our efforts, it's clear."

It's a savagely remarkable dimension to the story of monarchs that the wonder they provoke appears to be doing little to shield them from being rubbed out. The species has been the focus of intense grassroots action to tally numbers, plant milkweed, and resurrect lost habitat. It has been agonized over in the US Congress and is the subject of innumerable fundraising drives. Hundreds of mayors, from Quebec, in Canada, to Houston, in Texas, to Guanajuato, in Mexico, have signed a solemn pledge to boost monarch numbers by safeguarding their meadows, cutting down on dangerous chemicals, and educating the public about the importance of their tiger-colored visitors.

Yet despite all these angst-ridden efforts, the monarch still faces being erased from all but a few pockets of North America. This is a desperately sobering scenario for the thousands of other butterfly and moth species that share familial and habitual bonds with monarchs yet garner just a tiny fraction of the adulation, conservation funding, or even recognition.

Lepidoptera, which includes butterflies and moths, is the second largest order in the insect realm, comprising more than 160,000 named species. The true number, obscured by uncertainties and the undiscovered, is likely to be more than double this total. Because butterflies are colorful, conspicuous, and delightful, researchers and enthused volunteers around the world have keenly compiled population records of them that are unusually dense compared with other insect monitoring. We have little understanding of the overall health of Singapore's insects, for example, but we know that nearly half of the city-state's native butterfly species have disappeared over the past 160 years, probably due to the loss of vegetation.

The picture on butterflies is also fairly clear in Japan, where an analysis of 192 woodland sites by the government and a conservation charity found that 40 percent of common butterfly species declined in number between 2005 and 2017 and are likely endangered. The great purple emperor, an imperious insect enshrined as Japan's national butterfly, has suffered a 90 percent drop. The Japanese government has blamed damage by deer to vegetation, as well as pesticide use and water pollution, for this slump.

Data on insects in New Zealand are threadbare in most places, but a survey in 2019 found that half of Kiwis saw virtually no eggs, caterpillars, or chrysalides belonging to monarch butterflies. (Monarchs are found not just in the Americas, but in Australia, New Zealand, and some Pacific islands.) Across the Tasman Sea, butterflies are struggling in tropical northern Australia, one of the most biodiverse enclaves left on the planet. The Australian Butterfly Sanctuary, situated north of Cairns, is a lure for visitors from around the world who gawk at the 1,500 species of tropical and subtropical butterflies on display. The sanctuary had success for years in breeding around twenty species of butterflies and moths in a process that then suddenly faced complete breakdown.

The life cycle of a butterfly starts when a female butterfly deposits eggs onto a specific plant, carefully selected, since caterpillars are fussy eaters. Depending on the species and season, about two to ten days after an egg is laid the caterpillar nibbles its way out of the shell before turning into a conveyor belt of eating, munching through the leaves selected by its parent. Before long, the caterpillar becomes too large for its own skin and will shed it several times to allow further growth. Each stage is called an "instar."

In the final stages, a caterpillar will form a silk chrysalis, or cocoon for a moth, within which its body deconstructs and re-forms as a butterfly. Within four weeks, in tropical climes, this metamorphosis will usually be complete and the butterfly is ready to emerge. Some species will take longer, up to a couple of years.

Staff at the Australian Butterfly Sanctuary do all they can to support this breeding by growing food plants for caterpillars and stocking the place with plants favored by butterflies, such as pentas and ixoras.

Space is provided for the animals to flutter around as well as shade from the sun. The strongest caterpillars are cultivated to bring the success rate of egg to butterfly as high as 90 percent, compared with around 1 percent in nature.

Then, in 2014, abnormal things started to happen. The wet season's monsoonal rains were abbreviated, and then, bizarrely, Hercules moths started to emerge in the middle of winter. Another batch of caterpillars started to develop, but then started to pupate at the wrong instar, one too early, and died. The same happened to the following two batches of the season. It was carnage.

No one at the sanctuary could recall anything like this happening before. "It didn't make any sense," says Tina Kupke, breeding laboratory supervisor at the sanctuary. "That was the first time we saw the complete death of everything." Then, in August 2015, some orchard swallowtail butterflies emerged with a slightly curled upper wing and were unable to fly. Within a couple of weeks, all the caterpillars were not growing properly, and every one of the orchard swallowtails died.

Shortly afterward, the caterpillars of the Ulysses butterflies, the iridescent blue jewel in the crown, started to die off halfway through their life cycle as a caterpillar. "They stopped growing and they literally dissolved," says Kupke. "They liquidized. They literally disappeared." As the sanctuary held most of the breeding stock of Ulysses butterflies, eggs would be routinely sent in the mail to other permitted breeders. The problems at the sanctuary therefore caused the broader breeding population of the species to quickly collapse.

The following summer, the wet season didn't arrive at all, Kupke says, and instead the tropics were baked in unusually hot temperatures. Three more species collapsed. Staff frantically tried to find new breeding methods. Petri dishes, used for twenty-five years to raise the caterpillars, started to overheat or become too humid and could no longer be used. Caterpillars refused food plants that had been happily devoured for the past twenty-five years. Half of the sanctuary's species were lost within a couple of months, dying as caterpillars in various stages.

It was as if a curse had been cast. "For twenty-five years we had little hiccups but never had this sort of 100 percent wipeout, one after another," Kupke says. "Nearly every species went through some extreme trauma within a year and a half. It was something very, very odd." In their attempts to resurrect the Ulysses, sanctuary staff got desperate. They scrubbed everything clean in case some disease was lurking somewhere, then gave the breeding efforts a break for a few months in hope of a reset. Kupke even got a special permit to net wild Ulysses butterflies to see if the offspring of the wild population would fare better. They were raised both inside and outside the laboratory, using brand new sterile equipment. It didn't work. "It was heartbreaking," Kupke says. "In the last five years it's been a race to survive."

A more normal wet season and new breeding methods helped restore many of the species, but not to the levels prior to the die-off, and nothing seemed to work in reviving the Ulyssess. DNA testing was done on eggs and caterpillars, with the species also checked for various diseases. Nothing conclusive was found. It left open many theories: Had changes in temperature and rainfall caused this blight? Had the food plants somehow transformed? Had a toxin played a role? Was it a shift in the broader landscape? Whatever the cause, Kupke has heard similar tales from butterfly specialists across North and South America as well as Europe. "Everyone will tell you the same story," Kupke says. "It's not just us. I mean, thirty-two years doing something isn't long if you look at the age of planet Earth. But to see the impact over the last five years has been massive and dramatic. That's what I can say."

In Europe, the situation is particularly stark. Populations of grassland butterfly species plummeted by nearly 50 percent between 1990 and 2011, a crisis the European Environment Agency has blamed on the intensification of farming and pesticide use, which has made the land across the continent "almost sterile" for butterflies.

Eight key butterfly species found in grasslands across the European Union have suffered declines, including the common blue, a generalist found across Europe, parts of Asia, and North Africa, and the small heath butterfly, a creature with rusty red patches on its wings that is known for its particularly territorial males.

Butterflies are now reduced to such cramped ranges that in swaths of Europe they are found only on the odd grass verge next to a road or in the occasional railway siding. A fortunate few may find a home in a nature reserve, but not enough to offset the impact of vast tracts of vanishing grasslands. "If we fail to maintain these habitats we could lose many of these species forever," warned Hans Bruyninckx, the Belgian political scientist who heads the European Environment Agency.

The picture is barely brighter in the United States. A 2019 analysis of two decades of data on eighty-one butterfly species found in Ohio revealed that total abundance is dropping by 2 percent a year, a decline that means a third of the creatures have vanished in the state within less than a generation. "It's shocking to see the decline over twenty years," says Tyson Wepprich, author of the study. Wepprich suspects that other categories of insects are suffering a similar sort of decline; we just know more about the butterflies. "At the moment, butterflies are providing a surrogate for species we don't have good monitoring data for."

The citizen science approach used in the Ohio study—relying on the repeated observations of an army of dedicated butterfly-friendly volunteers—has its deepest roots in the United Kingdom. The famous J. B. S. Haldane quote that the natural world shows that God has "an inordinate fondness for beetles" would probably include butterflies if the British had a say in the matter. The oldest known pinned insect, still pinioned with its original fixture, is a Bath white butterfly, a rare visitor to the United Kingdom that was scooped up in Cambridgeshire in May 1702. Its abdomen is slightly warped and the stark whites and blacks on its wings have faded, but it remains a remarkable sight to those able to make the appointment-only viewing of it at the University of Oxford.

Butterfly collecting grew from the hobby of a few overseas voyagers who pressed discovered insects between the pages of bound volumes to become a leading passion of wealthy Britons. Specimens of species such as the large copper, which has since vanished from the country, fetched hundreds of pounds at auction. In Victorian Britain, some lepidopterists even became society celebrities, such as Margaret Fountaine, who collected butterflies throughout Europe, South Africa, India, and Australia.

Fountaine published numerous research papers and raised thousands of butterflies from eggs and caterpillars, and her illustrated sketch books of butterfly life cycles were deemed of sufficient merit to be housed at the Natural History Museum. In 2019, an unofficial blue plaque was erected in Fountaine's native Norwich—inscribed under her name was "I'm a bloody lepidopterist and I loved love."

Later, Winston Churchill became enamored with butterflies while he was a young man in India, subsequently creating a butterfly house at his red-brick estate in Chartwell, Kent. His wife Clementine planted buddleia, lavender, and other nectar-rich plants to encourage the butterflies that Churchill raised from caterpillars. Churchill, known for his "black dog" bouts of depression, found himself enchanted by butterflies and their metamorphosis. It was an era when nature was reveled in, and plundered, with little guilt: a legitimate hobby for children and adults alike would be to gleefully sweep the countryside with butterfly nets and compare catches, in a sort of terrestrial fly-fishing. This infatuation wasn't unique to the United Kingdom's leading figures. For example, the novelist and poet Vladimir Nabokov, author of *Lolita*, was consumed by an interest in entomology while growing up in the Russian city of St. Petersburg. He later organized the butterfly collection for the zoology museum at Harvard University, an institution that still holds Nabokov's "genitalia cabinet," where the author kept the accumulated organs of male blue butterflies.

The netting of butterflies diminished as conservation concerns grew, but the fascination with the colorful beings that flap bouncily around us remained. Groups of butterfly enthusiasts formed, trekking through forests and heath to marvel at unexpected finds. The British progressed from killing and impaling butterflies to counting them. These observations, gained through jotted sightings in a patchwork of areas called transects, have been corralled to provide the sort of data most entomologists would commit a minor crime for.

The recent trends make for painful reading. Since 1976, habitat-specialist butterflies—those that depend on particular landscapes such as heathland or chalkland—have declined by 68 percent, according to British government figures. Butterflies that are more generalist in their habitat tastes have dropped by around a third. A four-decade-long annual butterfly count, called the UK Butterfly Monitoring Scheme, shows that seven of the ten worst years for butterflies have occurred this century.

A more infrequent compendium released in 2015, dubbed the "state of the nation" for butterflies, declared a "serious, long-term and ongoing decline of UK butterflies," with 70 percent of species decreasing in occurrence and 57 percent dropping in abundance since 1976. Overall, three-quarters of the nation's resident and migrant butterfly species have been sighted less frequently or suffered population falls in this time. "The broad brush result to explain to someone in the pub is that three-quarters of British butterfly species have declined since the 1970s and a quarter are doing well," says Richard Fox, associate director at Butterfly Conservation, a UK charity.

Concerns over butterflies can be traced to 1979 with the UK extinction of the large blue butterfly, regal-looking light blue creatures that begin life as larvae in nests of red ants, feeding on its hosts. Determined attempts have been made to reintroduce this butterfly to southwest England, a quest that has not been without its complications. In 2017, Phillip Cullen, an amateur entomologist and former bodybuilder, was handed a six-month suspended prison sentence after vaulting a padlocked gate to enter a nature reserve in the Cotswolds, where he spent several hours swiping a butterfly net around. Police later found a large number of mounted butterflies, including two large blues, in Cullen's home.

Significant cash can still be made selling mounted rare butterflies as if they were Victorian artifacts. The age of insect scarcity is inspiring a new generation of criminals armed with nets, sensing the profit of demand as keenly as those who are behind the heist of beehives in California.

More widespread alarm over British butterflies didn't take hold until a major analysis showing declines came out in 2001, an effort that resulted in a decent amount of media exposure and even a question in Parliament. Some species have since rallied: the distribution of the flamboyantly named Duke of Burgundy butterfly shrank by 84 percent following the 1970s, but has since staged a major comeback in Sussex, Kent, and North Yorkshire following an effort to link up remnants of the duke's favorite habitat of grazed grassland and scrub. Still, all the surveys and noise made by volunteers and campaigners have, as with the monarchs, failed to arrest some major declines across a worryingly long list of butterflies.

Britain is now home to some vanishingly rare butterflies, such as the high brown fritillary, found in just a few isolated spots around the country. But, as Fox points out, even "bog standard garden butterflies" are now struggling. The large white, once so numerous it was considered a pest, experienced a 19 percent slump in abundance in 2017.

The woes of butterflies have dovetailed with those of moths. A major 2013 study of 337 common UK moth species showed that over a forty-year period until 2007, two-thirds suffered population declines. Larger moths are faring particularly badly, with southern Britain virtually a moth graveyard—a 40 percent decrease in total abundance was recorded for the region.

These losses can seem counterintuitive; the British Isles occupies a cool, damp corner of northwestern Europe that will become more popular with certain butterfly species as countries to the south increasingly become intolerably hot. The first decade of the twenty-first century also saw the United Kingdom double its spending on conservation measures. And yet, as a 2015 study pointed out, the total abundance of widespread butterfly species declined by 58 percent on farmed land between 2000 and 2009. This suggests that blame should be heaped upon neonicotinoids, particularly as butterfly populations are stable in Scotland, where use of such chemicals is relatively rare.

In the best-studied country for butterflies in the world, a nation populated by devotees of these insects, scores of species are still fluttering toward extinction. More data may salve the deficit of knowledge on insects, but when provided, it provides little comfort. Rather than peeking behind the curtain at a few bloodstains, UK researchers have yanked aside the drapes to reveal a sort of silent carnage.

Zoom out further and the detailed record keeping on British butterflies suggests that the same horrors lurk unquantified elsewhere. Many other countries, after all, have also cleared butterfly habitat, sprayed pesticides, and caused nitrogen pollution via the burning of fossil fuels, a process that turns soils acidic and swamps butterfly ranges with an unhelpful tangle of new vegetation. "There are common butterflies and moths, things you expected to see in a garden on a sunny day, that we are now going a few years without seeing," Fox says. "I'm 50 years old so I'm too young to remember meadows heaving with butterflies, but you just don't see many in the countryside at all now. Whenever I see one I think "oh good, I should note that." Fox worries that this lost Eden will slip from memories, blunting the urgency of restoration. "We need to keep reminding people what it was like," he says.

Few people have examined the sliding fortunes of butterflies more closely than Art Shapiro, a person who has hoarded a wealth of information on an array of things—comic strips, quotes, the politics of Argentina—but whose specialty is the butterflies of northern California. Since 1972, Shapiro has mounted a sort of one-man response to the British fastidiousness in butterfly research, repeatedly traipsing along the same stretches of the Sacramento River delta, through the Sacramento Valley, and up into the soaring Sierra Nevada mountains to scribble notes on what he finds.

With his voluminous beard and a shock of unruly hair that shrouds much of his face, Shapiro has long cut a distinctive figure on campus at University of California Davis, where he is a professor of evolution and ecology. The university is a 90-minute drive north of the Jasper Ridge Preserve, where another veteran biologist, Paul Ehrlich, started studying checkerspot butterflies in 1960, only for them to vanish by 2000. Originally intended as a five-year project, Shapiro's work has spooled out to become the longest continual insect monitoring program in North America.

On a mild morning in January 2020, I was in California to act as chauffeur and companion to Shapiro as we drove to one of his ten fixed research sites—a mix of open and wooded habitat along a stretch of the American River that buffers Rancho Cordova, a community of prim housing east of Sacramento. Back in the 1970s, we would likely see five or six butterfly species along this 45-kilometer (28-mile) strip of riparian terrain, Shapiro says. Today, if lucky, we'd spot one or two.

Shapiro has a low-tech approach to record keeping, his shirt pocket stuffed with blank cards along with two pens and a Sharpie. He will jot down the butterflies he sees, along with notes on the weather and vegetation, which colleagues back at UC Davis will feed into a computer. He declines to carry around a cell phone and, in a Californian rarity, doesn't drive a car.

Theoretically, this wedge of land in Rancho Cordova could yield a mourning cloak, known as a Camberwell beauty in the United Kingdom, a highly distinctive butterfly with maroon wings tipped in yellow. A red admiral is possible, too, as is a buckeye, famous for the eye-like pattern that adorns its wings. There have been a lot of "zero" days marked on the cards recently, though. "This winter has been terrible, absolutely awful," Shapiro mutters.

Shapiro has an enviable research location. He has seen more than 150 species of butterfly from sea level to the Sierra Nevada tree line, a bounty that could only be rivaled if he spent his time in select spots of the Alps, the Rockies, or the tropics. Butterfly numbers tend to yo-yo between the years—California's many microclimates help create a lot of noise amid the long-term signal—but Shapiro noted a gradual decline in what he was finding until the late 1990s.

Then came a crash.

"The break was in 1998 and 1999 when seventeen species at low elevation went abruptly downhill," he says. "That was the alarm bell for us that something serious was happening." A team of twelve scientists, including Shapiro, concluded that the dip was likely due to a shift in insecticide use toward neonicotinoids by northern California landowners, a deadly choice that caught butterflies in the cross fire.

Some butterflies gradually recovered, while others have faded from view. The large marble, its wings laced with whites and greens, was common in the 1980s and is now regionally extinct, as is the field crescent, its wings a riot of oranges, browns, and whites. The common sootywing, which once bred outside Shapiro's laboratory, is now erroneously named, with just one active colony remaining on Shapiro's transect. "It's sort of depressing but if I thought about it too much I'd commit suicide," Shapiro says. "Everything is now pretty much in the toilet."

The searing California drought that began in 2011, the worst in more than 1,000 years, was surprisingly beneficial to some lowland butterflies, although those at higher elevations, which rely on a sturdy snowpack to insulate them from freezing and desiccation, suffered badly. The easing of the drought was a relief to Californians, but restarted the downward trend of the butterflies, including monarchs. The most famed of all the butterflies enjoyed a huge boom in numbers during the drought but have since toppled from millions in number to the tens of thousands. It's unclear why, although warm, wet years can help unleash various bacterial and fungal diseases in butterflies.

"Everyone who planted milkweed was like "hooray the monarchs are saved!" even though that had nothing to do with it," Shapiro says. "Then they dropped again." The year 2018 was perhaps the nadir in his forty-eight years of study, the only time declines occurred at all elevations. "It was ghastly," Shapiro recalls.

Species like the pine white and the Ivallda Arctic, found only on the 2,774-meter (9,100-foot) Castle Peak mountain, weren't seen at all. The Ivallda Arctic was missing for three years in a row before a sighting in 2019. "We saw one last year. One! So it wasn't extinguished, yet," Shapiro says. His research walks are increasingly becoming a lonely exercise in grasping at diminishing delights. "It's like the whole universe is conspiring to destroy butterflies," Shapiro says. "I feel like a physician who has had a patient since he was a little kid, I know him well. The patient is dying, and I know he's dying. I just don't know why."

We cut across some grassland, Shapiro pointing out some fiddle-neck that has yet to bloom. Usually there is an expanse of yellow mustard on the Sierra foothills by this time, too, but everything is late to arrive in 2020, an odd break from climate change's relentless nudging of spring to appear earlier in the year. For the past four decades, Shapiro has offered a pitcher of beer to the first person who can bring him a live, adult cabbage butterfly each year after the winter dormancy. As the globe has heated up, the average first flight date of the butterfly has sped forward to January 18—about twenty days earlier than when he started.

An hour in and we have spied a couple of birds and a squirrel, but no butterflies. We switch back to follow the contours of the American River, trudging across the covered-up tailings of the dredging that ground to a halt in the wake of California's great gold rush.

Back in 1972, Shapiro had no inkling of the rolling doom of climate change or that his work would take him to Argentina to see his favorite butterfly of all—the silver butterfly of Patagonia, witnessed dancing in shafts of light that sliced through the clouds. Shapiro's plodded routes have stayed the same even if the butterfly world morphs and collapses around him. He is left with no other goal but to keep going, to keep walking 24 kilometers (15 miles) a day looking for butterflies. Shapiro is 74 years old and in generally good shape, although he can't visit Castle Peak anymore because although his knees will allow him to ascend, these days they won't easily allow the descent. "The objective is the same as my objective, which is to keep alive as long as possible," he says, rolling out an anecdote of how his aunt Minnie died of a heart attack while watching her favorite TV show, a soap opera called *Young Doctor Malone*. "If I am to go while looking for butterflies, that would be ideal."

Ultimately, what would we lose if all the butterflies were to withdraw from this world? A shared ancestor of moths and butterflies has been traced back to around 300 million years ago, with the intervening years featuring a slow-moving evolutionary arms race between plants and hungry caterpillars to the point of codependency.

The benefits are weighted rather more heavily in the butterfly's favor, however—no plant is completely dependent on them for pollination, while no animal would starve if they had no butterflies to eat. Cynically, for all the efforts made to save butterflies, our lives would go on unperturbed without them. "They are ecologically pointless," says Erica McAlister, who points out that flies may lack the popularity of butterflies, but are far more valuable as pollinators. "What really annoys me is that their larvae go around eating everything and yet we forgive them when they are adults, because they are pretty."

Butterflies' defenders may quibble with this assessment, but the argument is almost moot. If other insects help keep us alive, butterflies are surely one of the things it's worth keeping alive for. As the quote misattributed to Churchill, the butterfly afficionado, put it when wartime spending cuts to the arts were proposed: "Then, what are we fighting for?"

Shapiro started looking for butterflies as a 10-year-old desperate to escape an unhappy home on the outskirts of Philadelphia in the 1950s. He spent hours in woodland and grassland near his house, a copy of *Field Book of Insects*, by Frank Lutz, stuffed in his jacket pocket. Butterflies allow us, after all, to touch upon a hedonism of nature that elevates us from our surroundings. They are important as sensitive indicators of environmental change, but also, more fundamentally, as inspiring embodiments of serene awe, a magical tonic for our mental health. These insects are treasures to which no monetary value could realistically be ascribed. The insect crisis is not only ripping up the floorboards of our shared house but also stripping bare the beautiful art from the walls. "Basically it's a matter of aesthetics and sentimentality," says Shapiro. "If all the butterflies in the world went extinct tomorrow, no ecosystem would collapse. But people like them. They are pretty and they are harmless. I've met people afraid of butterflies, but that's rare."

After a 2-hour loop, we are heading back toward the car through a tangle of blackberries when Shapiro lets out a yelp. A startled passerby looks around in alarm to see this extravagantly hirsute man waving an arm at a passing moth, known commonly as a highflier. I saw nothing. The zero day has been saved, just. There were no butterflies, the closest being a solo monarch that inexplicably fluttered over US Route 50 on our way to Rancho Cordova.

"I'm disappointed to not have seen anything but that's the difference between science and art," Shapiro says. "Art transcends reality while science depicts it."

Back in central Mexico, the butterflies are more than just adornments. They are an economic engine for an area plagued by poverty and occasional bouts of violence. But their beauty has long been appreciated, too: pottery from before the Spanish conquest has been unearthed, emblazoned with images of butterflies. Depictions of monarchs are found on car license plates, while schools, football teams, and businesses are named after them. Sáenz-Romero says his plan to move the forest up the mountain is starting to gain traction with a few government officials, although the pace of progress is far too slow.

A popular Chinese proverb states: "The best time to plant a tree was twenty years ago. The second best time is now." In the case of migrating monarch butterflies, not much lies beyond "now." If mature, 80-year-old oyamel firs are to be in place for the timetable set by climate change, they would ideally be well established by now, at high elevations, in their tens of thousands. There's no indication that this scenario will happen at any scale. The butterflies will likely vanish from here, as they will elsewhere. "We need to plant these trees and do it now, massively," Sáenz-Romero says. "But that's not being done. Maybe it's a dream to save the butterflies. But we have to do something."

To traverse the winding, rutted track taking you into the Sierra

Chincua sanctuary, at the heart of the butterfly reserve, you have the choice of walking or taking one of the assembled ranks of horses on offer. Sáenz-Romero and I, having left the *ejido* behind, take the equine option and slowly lollop through the lofty city of alpine firs, the milieu more *après-ski* Switzerland than Mexico.

The final portion is negotiated on foot, with a few introductory monarch butterflies swooping overhead as we approach a rocky outcrop at an altitude of 3,150 meters (2 miles), a cozy middle ground in the oyamels' preferred range. At the summit of the outcrop a glade of monarch-laden trees stand in a semicircle as the terrain plunges downward again, as if the butterflies chose the spot for the sweeping view of the string of forested volcanoes that dot Mexico's core.

Millions of monarchs envelop the firs, a blur of orange obscuring the green of the needles. Clumps of butterflies coat the branches while others sun themselves on the rocky ground. Some nourish themselves on nearby plants. Then, as if in a waking dream, the wind stirs and ripples of butterflies take to the air, floating upward and darting around the trees at sharp angles. There's an audible murmur from a dozen onlookers, interrupted only by a woman who is convinced she has a wasp tangled in her hair. Other than that, the only sound is the patter of the monarchs' wings, like light rainfall on a canvas tent. It's a transcendent moment.

On the way out of the sanctuary, local vendors are selling any sort of product—pens, hats, baskets—that could possibly feature the depiction of a monarch butterfly. At the exit, there is a pair of stone butterflies, a sculpted monument to Lincoln Brower, the world's most celebrated monarch expert, who was a regular visitor to see the migration in Mexico before passing away in 2018 at age 86. Brower took Jimmy Carter to the butterfly sanctuary in 2013, with the former US president peppering him with questions. On the way out, the two men noted that much of the vegetation had been trampled on by tourists as they eyed the dozens of tour buses parked near the visitor center. Two years later, Brower was a signatory to a petition calling for the US government to list the monarchs as a threatened species. He died before seeing whether the animals he had devoted four decades to would be officially protected.

"We need to develop a cultural appreciation of wildlife that's equivalent to art and music," Brower said, in one of his final interviews. "We should care about monarchs like we care about the Mona Lisa, or the beauty of Mozart's music."

8

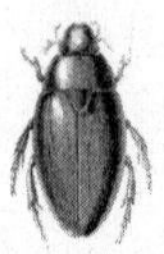

The Inaction Plan

The most visible sign of alarm over the insect crisis has perhaps been the number of people now prepared to earnestly wear bee costumes while attempting to convey controlled outrage. In the chill of 2019's fading winter, there were plenty of bee-clad humans to be seen in southern Germany, frowning through black and yellow face paint as they waddled around with their bulbous abdomens.

There have been other, similar protests since the honeybee was elevated to its rather misleading status as an environmental avatar, and onlooking citizens in Bavaria could be forgiven for dismissing the throng brandishing a mixture of English and German placards—"Bee a hero!" and "Nur mit uns brummt die wirtschaft" ("The economy only hums with us")—as just another group of fringe eccentrics.

But this time a minor political earthquake was stirring. A referendum on biodiversity had been put forward by a coalition of conservation and political groups calling for life in the farming heartland of Europe to be upended in order to rescue insect life. The petitioners called for 30 percent of farmland to become organic and insect-friendly, for wetlands and hedgerows to be restored, for pesticide use to be slashed, and for light pollution to be curbed. A sort of charter for insect renewal, the Save the Bees campaign seemed daring, transformative, and, above all, utterly hopeless.

Bavaria is not just the most conservative state in Germany but also the most fiercely pro-farming. Agricultural muscle is showcased on a grand scale here, with sweeping fields of monoculture crops vigorously treated with pesticides. Previous attempts to enact environmental restrictions or even to erect wind turbines have been stymied. The conservative state government, therefore, felt confident to laugh off this strange new grassroots effort to handicap farmers just to aid some creepy-crawlies.

In the end, though, it wasn't even close. In a stunning example of direct democracy, the petitioners secured the support of 1.75 million Bavarians, a fifth of the electorate and well beyond the 10 percent threshold required for the state government to enact the proposals. Two years after the deathly Krefeld insect surveys, voters looked around at surroundings shorn of crickets, butterflies, bumblebees, and skylarks and decided they'd had enough. "To be honest we weren't that optimistic at the start," says Markus Erlwein, who became a main spokesperson for the Save the Bees campaign. Then he saw long lines of Bavarians brave frigid temperatures to file into town halls and sign the petition as the media trumpeted tales of insect doom. "The ground was set, it was the right time. It just exploded," Erlwein says. "When the result came in, I cried."

There was no immediate fairy-tale ending, of course. The powers of big agriculture mobilized against the rumblings of change, and Bavaria's attempt to switch to wilder, more pesticide-free methods has collided awkwardly with the European Union's policy of shoveling money at farmers to maintain the status quo. But the continent that pioneered and exported methods worldwide that have brutalized insects and the towering edifice of life that depends on them is starting to realize that the tools of modernity, as the naturalist Aldo Leopold put it, "suffice to crack the atom, to command the tides. But they do not suffice for the oldest task in human history, to live on a piece of land without spoiling it."

Slowly, belatedly, this skein of reckless progress is being wound back—France has banned all neonicotinoid pesticides, Germany has sought to shut down bright lights after dusk, Norway has created a safe haven for bees in the heart of Oslo. The response to the insect crisis has been piecemeal, underfunded, and occasionally muddled, but the outlines of restoration are there. Insects are now, at the very least, being included in the gloomy conversations about the biota that are being stripped from our planet. "A few years ago any insect not a bee or a butterfly would be considered a pest," says Josef Reichholf, a German biologist. "That has changed. We have a better feeling with the word 'insect.' We know honeybees are important but now that others are, too."

Reversing our destruction of insects can seem a complex challenge, involving the overhaul of a gargantuan agricultural machine, the evolution of cultural norms, and the decoupling of improved living standards from environmental annihilation. "The threats don't add, they multiply," says Pedro Cardoso, the Finnish Museum of Natural History biologist. "If species have to endure only one of these threats, maybe they could survive. Two or more threats and it becomes a big, big issue." Our tenuous regard for insects means that these reforms would have to be primarily driven by other motivations (our own health, possibly, or action on climate change), which would hopefully align with insect conservation.

But if you squint a little, addressing the insect crisis can be viewed as surprisingly straightforward. In essence, we would just stop doing certain things. The mere absence of action, of letting things slide a little, could be enough. We don't have to build and launch the equivalent of *Apollo 11* or design a whole new clean energy grid or race to find a vaccine for a pandemic that has paralyzed the world. The task, fortunately, involves far more indolence than that.

The more contentious parts of this plan involve restricting the use of certain chemicals and allowing insects room to repopulate by giving over land, even at the margins, for a riotous selection of wild plants. But much of the renewal is far more domestically humdrum, such as reducing the number of times we mow or weed our lawns or deciding not to use dazzling outdoor lights. A step further would involve considering whether you need an ordered, manicured lawn at all. The entomologist May Berenbaum sums up this approach as "not an action plan but an inaction plan." We've fussily reshaped our environment to the point of breakdown. Perhaps it's time to sit back and see what could blossom in front of us if we just give it the chance.

What would this look like? Some environmentalists point to the example of an ecological revolution quietly underway in a pocket of southeast England. Knepp, a 1,416-hectare (3,500-acre) farm in one of the most densely populated corners of Europe, is a window to a world where intensive farming hasn't cut down insects and myriad other creatures. In many respects, Knepp isn't really a farm at all, having been given over to an ambitious and far-reaching project to allow nature to take over and shape the land with minimal human meddling.

Knepp's owners, Charlie Burrell and Isabella Tree, did try for years to run an orthodox arable farm by investing in heavy machinery and pesticides, struggling to grow profitable crops on the clay and limestone of the Lower Weald that makes the ground rock-hard in summer and sludge-like in winter. Much like the way the native people of the Arctic have dozens of different words for snow, locals in Sussex have more than thirty dialect words for mud. The challenging conditions and might of larger industrial farm competitors drove up Knepp's debts. In 2000, the decision was made to sell the farm machinery and dairy herds to save the estate from financial ruin.

Then, a moment of epiphany—an area of the farm was given conservation funding to be restored from cropland, and almost immediately wildlife moved in, with insects in the vanguard. "There was this buzzing, thrumming, heaving feeling of having insects all around us," Tree recalls. She would walk through knee-high native grasses and daises, kicking up grasshoppers with each stride. "It suddenly felt completely different," she says. "The scales fell from our eyes." Inspired by the work of Dutch ecologist Frans Vera, Burrell and Tree decided to forgo crops across the whole farm in favor of free-roaming herbivores that would influence the ecosystem through their grazing. The land is free from pesticides, and the animals, which include Exmoor ponies, Tamworth pigs, and 400 head of cattle, aren't given antibiotics or other interventions. Knepp somehow makes money from this arrangement, via the sale of 75 metric tons (83 US tons) of organic meat a year, along with camping for paying ecotourists and the renting out of former farm buildings. It turns out that the biggest risk Burrell and Tree could have made would have been to continue farming as before.

Thickets of scrub have shot up, and dead wood is allowed to lie rotting on the ground—untidy eyesores for your standard farmer but thrilling for insects. More than 600 invertebrate species have been recorded at Knepp, including the violet dor beetle, which hadn't been seen in Sussex for fifty years. More than twenty species of dung beetle have been found in a single cow pie, Tree says. Rare click beetles emerge from larvae found in the soft stumps of old oak trees, while mayflies and dragonflies flit around sparkling clean ponds and lakes. The scarce chaser, a dragonfly with blue eyes found only in a handful of places in the United Kingdom, flourishes here. The largest British population of the purple emperor butterfly is also found at Knepp, leading to hordes of aficionados descending on the estate each summer brandishing rotting fish, smelly cheese, and dirty nappies as bait to lure out this elusive and idiosyncratic butterfly. "It's incredible how quickly wildlife comes back and insects are very much the first to come," says Tree.

The resurgence in insect life has most obviously benefited the avian family, with nationally threatened birds such as turtledoves and nightingales now regularly seen at Knepp. But there is another upside that should interest even the most unsentimental farmer: the proliferating dung beetles pull manure down into the earth, helping replenish the soil with nutrients. The insects help give structure to soils that are being lost, globally, at an astonishing rate. The United Nations has estimated that as much as 40 billion metric tons (44 billion US tons) of topsoil is lost through erosion worldwide each year due to a punishing

regimen of intensive cropping, plowing, and chemical application. "If you switch to nature-based farm management, including accepting or even encouraging insects, your profits can rise because there are no inputs, you are not destroying your soils, you are producing organic crops that can be sold for more," Tree says. "Restoring our soils helps solve almost every crisis facing us at the moment, including the climate problem. The question is whether people can realize this quickly enough to prevent further damage."

Tree is adamant that all types of farmers, even those who don't fully own their land or have attractive glamping scenery around them, can drop their hands from the steering wheel and let nature intrude a little. This isn't rewilding in the best known sense of the word—Knepp hasn't reintroduced apex predators such as the wolves that successfully reshaped Yellowstone National Park in the United States in the 1990s or even creatures such as the beaver that has been brought back to parts of the United Kingdom in recent years—but it is a "wilding," as the title of Tree's book on Knepp alludes to. This is handing over a slice of control to nature, rather than a sort of strict picture-book restoration. "It's a common misconception that rewilding is bringing something back to what it was before," Tree says. "The environment has completely changed from even fifty years ago, you could never re-create the past. It's about reintroducing the tools of nature to create dynamisms in the present, to create a novel ecosystem."

A different, healthier paradigm seems tantalizingly within our reach. The twin disasters to have hit latter-day Britain—the coronavirus pandemic and Brexit—have assailed the country but also provided the unlikely side effect of offering a gentler, happier deal for insects. The pandemic saw the abandonment of much of the roadside trimming that keeps grass verges—vital insect habitat—closely shorn. A bounty of wildflowers sprung up, triggering a remarkable wildlife recovery—just one short stretch of roadside vegetation in Dorset suddenly contained half of all known butterfly species in Britain, including the small blue, the smallest butterfly in the country. Meanwhile, the United Kingdom's painful exit from the European Union

has provoked a huge upheaval in farming policy, with the bloc's system of subsidies for landowners replaced by national payments based on the replenishment of soils, cuts in pesticide use, and the expansion of woodland.

Ambitious thinking has started to sprout, such as calls for a quarter of Britain to be handed back to nature, allowing areas badly suited to crops like Knepp to again teem with insect and other life. Dedicated conservationists have even managed to start resuscitating seemingly hopeless situations, such as the loss of the short-haired bumblebee, which was once common across southern England but declared extinct in 2000 due to the endless mangling of its preferred grassland habitat. Over the past decade, dozens of queen bumblebees have been flown over from Sweden to reestablish the species in Dungeness, a headland of shingle beach and marsh in Kent. With just a few tweaks from partner landowners (such as moving livestock to new pastures and restoring wildflowers), a host of bumblebee species, not just the short-haired variety, have come roaring back.

But rewilding, or simply restoring, a scattering of places in our industrially scoured world won't be enough. As a major 2020 research paper on tackling the insect crisis warned, we must protect everything, from the cathedrals of rainforests to the scruffy incidentals of weedy railway sidings, but also ensure that these refuges for insects are joined up. "Conservation efforts have largely been focused on charismatic megafauna, especially birds and mammals, with little thought on ecosystem connectivity," the researchers wrote. Wildlife corridors have, until now, mostly meant building specialized bridges or underpasses to ensure the migrations of pronghorn in Yellowstone, reindeer in Sweden, or even red crabs on Christmas Island. Now this thinking is starting to include insects, too. Insects need safe corridors to move between suitable habitat in order to protect genetic diversity, find better food resources, and retreat from the relentless advance of climate change. Isolated reserves can only do so much if insects have to risk a deadly gauntlet of chemicals and concrete in order to get there.

The idea of providing space on cropland for insects was until recently

so outlandish that a proponent, scientist Stefanie Christmann, faced broad derision from agricultural researchers when she argued her case a decade ago. "They thought I was a crazy environmentalist. They just laughed at me," she recalls. The laughter has receded somewhat since then, the insect crisis coming at a time when farmers' understanding of the usefulness of pollinators has grown. The European Union has handed out money to farmers to create strips of wildflowers for many years now, but success has been intermittent given that landowners are being asked to go out and seed a jumble of vegetation they reflexively think of as weeds with no ongoing financial upside.

Christmann believes there is a far more farmer-friendly alternative that could become a blueprint for countries, especially those outside the European Union's pollinator plan. She has spent several years traipsing around fields from Uzbekistan to Morocco talking to farmers about a simple but revolutionary change—why not plant herbs, spices, or fruits in the borders and unused spaces at the fringe of fields? Cucumbers, sour cherries, strawberries—the farmer could choose the best option, but the outcome would largely be the same. A network of pollinator-friendly habitat latticed through landscapes that, crucially, provides an income for farmers. At first Christmann faced skepticism—why hadn't she bought hives of honeybees if she was bothered about pollination? When were these weedy things growing near the crops going to actually start making money? But within a year or so the landowners were delighted. "The farmers felt they were respected and part of the team," Christmann says. "For us they are the protagonists of pollinator protection."

Not only have these experiments brought wild bees, flies, wasps, and other pollinators buzzing back, but they have also resulted in reduced pest abundance, by as much as half in some cases. Separate research has shown that encouraging certain predatory insects can provide a natural shield against crop pests, negating the need to bombard crops with chemicals. Of particular promise is the use of parasitoids, typically wasps or flies that lay their eggs on or in another insect, ultimately killing them as the larvae grow. Universally rolling out Christmann's win-win arrangement will still take persuasion and funding, and there are certain landscapes, such as the US Midwest or California's Central Valley, that are so degraded that it will take some time before veins of insect life course through them again. But even in these places there is a budding will among farmers to do things differently, from planting a few stands of milkweed for monarch butterflies up to embracing the tenets of regenerative agriculture, where reduced tilling, a shift away from synthetic fertilizers and pesticides, and the planting of cover crops help improve soils and counteract erosion. "I know that agriculture will change, it's not a question of if but when," says the scientist turned no-till farmer Jon Lundgren, who has angst-ridden conversations with other American farmers enticed but wary to break from the orthodoxy. "We don't have a choice. What's it going to cost you not to change? It's going to cost your farm. It's going to cost your grandkids. Read the writing on the wall. The insect apocalypse is just the first sign of this."

*

IF WE PUSH the bounds of optimism further, we can contemplate comprehensive networks of national and international protected corridors, forging capillaries of thrumming insect activity through even biologically impoverished landscapes. Buglife, a British insect conservation group, has used computer models and on-the-ground verification to come up with a pioneering model of this, which the group calls B-Lines. The lines are insect pathways that weave throughout the United Kingdom's towns and countryside, looking on a map like a pile of red sauce spaghetti dropped from a height. The work is often painstaking, involving negotiations with different layers of government and landowners, but several thousand projects are already underway and the reception to the idea has largely been warm. Buglife's grand ambition is a total of 5,000 kilometers (3,000 miles) of insect corridors, each comprising wild flowering habitat around 3 kilometers (almost 2 miles) wide. But even a fraction of this would allow some imperiled insects an escape hatch from overheated, toxic, boxed-in homes.

The problem of habitat fragmentation has become so appalling that butterflies risk developing smaller wings and punier flight muscles because they are marooned for generations, says Buglife's chief executive, Matt Shardlow, who suspects that some of the decline in surveyed insects could be down to the simple fact that they aren't able to fly around as much. "By fragmenting the habitat and making the areas in between so inhospitable we are actually cutting back on evolution's chance to respond to the problem," he says. "As climate change comes along, we need to provide more stepping stones to help species move again." Insect conservationists like Shardlow spend a lot of time pondering localized threats to certain species, but increasingly the thinking has become grander—of transnational insect highways, of sweeping bans on chemicals, of revolutions in land use practices, of a new deal for our relationship with the insect world. What if, they wonder, we could call a total cease-fire in the war we've waged on insects, not just in our rural areas but everywhere, even deep within our cities?

*

IT'S HARDER TO IMAGINE a place less welcoming to many insects beyond, perhaps, the odd scuttling cockroach than the more spartan areas of New York City. Even with the rapid onset of gentrification, swaths of the borough of Brooklyn are dominated by utilitarian concrete and metal, hues of beige and gray punctuated only by the odd London plane tree. Just like New Yorker rats that have developed a taste for pizza, insects have adapted to this place in unusual ways. A few years ago, amateur beekeepers who tended to hives in the city were startled to find honeycombs striped in shocking red rather than amber. It turned out the bees had been flying over to a nearby factory to gorge themselves on red food dye used in maraschino cherry juice.

The gritty heart of the city's heavy industry beats aside the Newtown Creek, a nearly 6.4-kilometer (4-mile) tributary of the East River that forms part of the border between Brooklyn and Queens. Once fed by freshwater streams that mingled with the briny East River, the area was an ecosystem of sprawling tidal salt marshes layered upon glacial deposits left by an ice sheet that retreated 12,000 years ago. This thriving ecosystem filled with fish and insect life was largely undisturbed until the surrounding area, now called Greenpoint, was taken by the settlers from Native Americans in exchange for just a few beads and axes. The wetlands were then drained and filled as agriculture took hold; then, in the mid-nineteenth century, the United States' first kerosene refinery and then first modern oil refinery were built here. By the end of the century, industrialists, including John D. Rockefeller's Standard Oil, had erected more than fifty processing plants on either side of the creek, which had been widened and deepened for shipping. Industrial activity from fertilizer manufacturing to sugar refining jostled alongside the hulking oil refineries.

As industrial waste was casually dumped into the waterway and surrounding lands, this small stretch of New York quickly became one of the most polluted and foul-smelling places in the world. Then, in 1978, a further disaster became apparent—a US Coast Guard patrol spotted a plume of oil spreading into the waterway, a calamity that resulted in at least 64 million liters (17 million gallons) of spilled petroleum products leaving a sickly sheen on the water, the creek's bed and banks daubed black. A remedial operation to extract a morass of waste from the bed of the creek and tentacles of contamination that have reached deep into surrounding soils will take many more years to complete. If a place could stand as a visual repudiation of the lofty ideals of the Knepp estate, this would probably be it.

Yet even in this bleak graveyard for insects, some space is being carved out for their resurrection in a fresh approach by government and conservation groups. A nature walk lined with swamp rose, woolgrass, and other plants that connect the area back to its marshy past is being threaded through the network of chunky industrial buildings that dominate the creek. The apotheosis is a former Standard Oil lubricants factory that has been given a startling hairpiece—a wildflower meadow installed on five different tiered roofs atop the building. It's a shockingly verdant dash of greenery in a landscape of brutally human domination that has provided an unexpected oasis for insects.

I took the G train to see the site, called Kingsland Wildflowers, in the summer of 2020, a period when a pandemic that had horrifically ravaged New York appeared to be waning a little. As I walked beside trucks rumbling to and from various factories, the scorching weather was a reminder why cities such as New York are attempting to usher back more green space: grasses and trees help take the edge off sweltering urban heat that is absorbed and radiated by concrete and tarmac. "Green infrastructure," as it is unromantically referred to by planners, can also improve air quality, bolster residents' mental health, and soak up stormwater that would otherwise cause sewerage systems to overflow into rivers and streams.

Kingsland Wildflowers sits atop a stout, red-brick structure that has been repurposed to house a film and TV production company. Passing the incongruity of a mobile Covid testing site, I walked up a couple of flights of solidly metal stairs to meet Lisa Bloodgood, of the community-based group Newtown Creek Alliance, who showed me a space that has become both an educational hub and a bulwark of ecological restoration. The aesthetic is remarkable—rooftops carpeted in a gloriously green cacophony of wild grasses and flowers that overlook a backdrop of scrap metal recycling, car repair shops, and other hives of labor. A digger loads debris into a waiting barge on the creek, a scene that has its own backdrop of a steely skyline of Manhattan skyscrapers slowly cooking in the sun. The trifecta of environments—urban wildflower enclave, rusting heavy industry, and gleaming behemoth of finance and the arts—is dizzying to see in such proximity to each other.

The roofs host around 9,000 hectares (22,000 acres) of native herbaceous perennials, including wild strawberry, goldenrod, and milkweed, alongside native grasses and shrubs. Much of the surface area is taken up by sedum, giving the tiered roofs a spongy appearance. Some green roofs are intensive, where soil depths are 15 centimeters (6 inches) or more to support trees and larger shrubs, while others are extensive, shallower soils for low-growing plants. Kingsland Wildflowers has both, matted on top of layers of membranes for drainage, root protection, and insulation. It's a remarkable biological reengineering of a place that could easily have been surrendered as yet another place unsuitable for wildlife.

"We get a lot of native bees here, a lot of butterflies, a lot of different kinds of wasps," says Bloodgood. Many of the solitary bees are able to dig into the soils to make a home for themselves. Congregating moths have attracted bats. There's been a surge in bird sightings, too—mockingbirds, swifts, hawks—as insect numbers multiplied. "This is one of my favorite areas," says Bloodgood as we look over a patch of black-eyed Susans basking in the heat as, beyond, New York's largest sewage plant looms, shaped like four metallic alien eggs. Bloodgood hops from one hot, exposed paved stone to another stone that is crowded by plants and remarks on the cooling effect of the vegetation.

After some time on the roof, it ceases to be freakishly abnormal. Instead, the gaze wanders to other, bare, building tops that now appear to be lacking. They seem oddly barren. We have drawn ordered lines on maps to demark boundaries for housing, for industry, for commerce, a bit for wildlife, as if our world wasn't a mutually dependent tangle of connections. We have squashed insects from our surroundings when we should have been inviting them in, for the health of things critical to our lives and the beauty they bring to even small pockets of ecologically plundered cities and farms. Belatedly, we are waking up to our foolishness—New York now requires new buildings to consider installing green roofs, Detroit is placing bee colonies in derelict areas, Munich is planting flowering strips that have attracted a third of local insect species within just a year, Utrecht is transforming bus shelters into bee sanctuaries. Perhaps, in ecologist Roel Van Klink's analogy, we are slowly learning how to lift our foot a little off the submerged log and allow insects to rebound. "We've done more than enough damage—the forests are gone, the salt marshes are gone, the meadows and grasslands are gone," laments

Bloodgood. "If we can just bring a little bit of that back, that is critical. This is the underbelly, the viscera, of the city. If we can build a thriving meadow on a building like this, in a place like Newtown Creek, then it can be done anywhere."

*

BUT JUST AS the sheer scale of the insect crisis is unclear, so is the capacity for recovery. Insects have a prodigious ability to reproduce and cling onto any environmental life raft pushed in their direction, but we have exterminated so many of them across landscapes that it's uncertain they will bounce back to their previous diversity. Even if there were a chemical-free, ecologically connected world led by climate activist politicians, some species we cherish still wouldn't be able to make it. "You can take the foot off the log but if the log has disappeared it's not going to float up anywhere," as Matt Shardlow puts it. "There's no guarantee species will come back."

There are so many intertwined threats facing insects that there is no simple escape for them. Even the utopian vision of farms transformed into Knepp-like cornucopias of nature becomes entangled with realities such as how to feed a ballooning global population while restoring land to low-impact, organic wildlife havens. Wealthy European countries may be able to reverse destructive agricultural practices, but if they continue to import a large amount of food, they still risk rainforest loss overseas as worldwide demand for farmland rises. The threats to the treasure chest of insect life found in the tropics, menaced by climate change and deforestation, "scares me to death," admits entomologist David Wagner. "We can't have no tilling farms across the planet if we have to feed 8 to 10 billion people," Wagner says. "We will need all sorts of farming. We need to dedicate certain areas of the planet for very intensive agriculture—more chemicals, more genetic engineering—so we can increase the yield per acre. What they are trying in Europe is well intentioned but it does make you scratch your head."

There was no stable state that has suddenly been wrecked, no nirvana for insects to return to. Even the grandest conservation idea of recent times—E. O. Wilson's "half Earth" concept, where half of the planet's surface becomes a human-free nature reserve in order to restore biodiversity—would fundamentally transform the world, rather than simply restore it. We are going to have to plot a new path through a radically different world, one that will continue to be warped by the climate crisis and the whims of governmental and lifestyle change. Restoration of habitat and cuts in pesticide use will be necessary, but these changes must be able to work within a broader societal scaffolding. We will have to keep adapting to our morphing environment but this time also ensure that we take insects along with us as we do it.

"The very concept of the Anthropocene is that we have moved into a new state as a planet, but it's not a static state," as Chris Thomas, the University of York biologist, puts it. We are on a continuum, like the notches on the thermometer between boiling and freezing, where the temperature has edged up and down skittishly for as long as we've been aware of the world around us. The ghastly difference now is that we are careening toward a reality out of the bounds of what has occurred for millions of years. "It's about our choices of speed, directions, types of change, not a discussion of change versus no change," says Thomas.

A grander perspective on how to navigate out of the crisis faced by insects, and humans, can be found in London's Natural History Museum, which commands a collection of 34 million insect specimens. The museum, a stately Victorian structure long famed for the skeleton of a diplodocus in its entrance hall (more recently replaced, in 2017, by a blue whale), suffers like other institutions from a legacy where collecting insects, especially common ones, was rare until several hundred years ago. Even since then, the field has long been dominated by hobbyists focusing on the most interesting or prettiest species. A lack of depth in time is compounded by a lack of depth in species populations—the museum has around half of all known fly species but most are made up of a single specimen for each, referred to as holotypes.

The collection still holds fascinating treasures, such as the Picasso moth, its wings like a canvas filled with geometric lines and shapes. There are stick insects that look just like leaves, blue-green jewel beetles with startling tufts of orange hair, and huge crimson butterflies that, when alive, like to gorge on rotting flesh and fruit. There's even a specimen of an ancient robber fly, which was caught in 1680 by the Queen's gardener at Hampton Court and kept squashed in the pages of a book.

But there has never been the means nor the will to thoroughly chart insect population trends—after all, what was the point? The sudden jolt of angst over insect declines has therefore left the museum frustratingly flat-footed for historical context. "Funding is our biggest problem. I've got to persuade you that a small black fly is worth as much as a panda," says Erica McAlister, the entomological curator. "So we get to this point where we realize things are going wrong with insects, the public suddenly notices and says 'where is all your data?' and we are like, 'Are you kidding me?' "

Climate scientists are able to read the runes of tree rings to ascertain previous eras of drought, or pierce the ice sheets of Greenland and Antarctica to extract towering, mile-long cylinders of ice that can tell them the temperature, atmospheric composition, or even wind patterns from hundreds of thousands of years ago. There are no such natural troves to tell us the past vagaries of insect populations, and even the best collections are somewhat haphazard. London's Natural History Museum, for example, has insects marked as being collected in places like "Surrey," "Yorkshire," or "Scotland," but then things get a little vague further afield—some specimens just cite "Africa." McAlister says that a personal favorite, apt for the Brexit age, is a butterfly simply labeled as "foreign."

So entomologists are increasingly leaning on advances in big data and genetics to plug some of the gaps. Museums from various countries are starting to digitize and share data on their entomological collections, to construct a sort of historical scaffolding around insects. Looking at genomes offers further insight. Subtle genetic changes in insects as they respond to factors in their environment can reveal population changes over time and area. Pollen fragments can be taken from bees and blood can be extracted from mosquitoes to see what they were feeding on, and when.

Extracting DNA from insects, in some instances, is the easy part. The danger is destroying specimens while doing this. The idea of lopping legs off rare or extinct insects is something that "fills us curators with fear and loathing," McAlister says. One of the ways she, along with specialists at the Wellcome Sanger Institute, a genomics and genetics research center based in Cambridgeshire, is trying to limit damage to the delicate specimens is to see if they can extract historic DNA by gently washing a few mosquitoes using an ethanol-based solution and a type of buffer that removes genetic residues from the specimens.

The insects are then put through a device called a critical point dryer, which not only leaves delicate specimens intact but also somewhat revives the appearance of corpses that have collapsed eyeballs, deformed wings, or shrunk abdomens. "We have basically set up a high-class hairdressing salon," says McAlister. "The Sanger Institute is washing the specimens and I'm blow-drying them afterwards. We are just glorified hairdressers with PhDs."

Despite all of these intricate efforts to describe which species are found where, the patchy records kept on insects stymies any attempt to conclude that they are now suffering a decline never previously seen on Earth. Yet scientists have the evidence that the current crisis is a deep and painful one and that many of the everyday practices we follow are unsustainable for large swaths of insect life. How the travails of the past compare with those of today are somewhat moot compared with far more urgent questions of where the declines are leading us and whether we can correct course in time.

Scientists tend to diverge on the question of whether this trajectory is one that will cause a series of cataclysmic tipping points, but a supplemental question helps pull the situation into sharper focus: tipping points for *whom*? Insects are being shifted to an unhappy state where there will be far more bedbugs and mosquitoes and far fewer bumblebees and monarch butterflies, but they will find a way to navigate future deviations on the continuum. As Thomas points out, while there is a net decline in insects, roughly a third of the species that cope well in a human-centric world are increasing, so the numbers aren't quite on a path to zero.

Humanity, on the other hand, may not be quite so resilient given the diminished mix of insects we have helped engineer. The grip of unease for people already food-insecure, for the health of our environment, and for the enveloping web of life that sustains and enchants us will only tighten unless we can rapidly reform our relationship with insects.

The outlook even spooks committed advocates of the more unloved insects that may proliferate in favorably altered circumstances. While McAlister will staunchly defend houseflies as being effective pollinators of bell peppers and other foodstuffs, she admits we probably don't want to help create hordes of them given that the flies can spread feces and bacteria via their legs as they walk around on various surfaces.

"I do have a self-preservation side," McAlister says. "I'd quite like us to survive but I'm worried we are going to see in our lifetime—actually, we are already seeing in our lifetime—really significant changes. . . . I think there's going to be a tipping point where we are going to have a big shock and be like 'oh God, we haven't mitigated against that, that and the other.' The current state is not good. It's really not good and we have to act quickly."

9

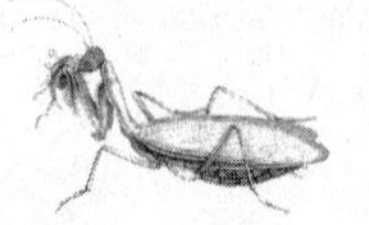

A Human Emergency

What, though, if we don't act quickly enough? If the fall of insects' tiny empires causes whole ecosystems to unravel, toppling previously solid certainties about the way our world functions, what then?

It's easy to foresee how diminishing supplies of certain foods and crashing wildlife populations will heap cascading suffering on the poor and vulnerable, given the lopsided nature of societies, and perhaps even stoke embers of resentment and nationalism as foundational resources become scarcer. It's also reasonable to anticipate that we will reflexively grasp for a technological fix to the mess we've created. Our zeal for quick, shortcut solutions to self-inflicted problems has fueled confidence, among some, that climate change can be solved by huge machines sucking carbon dioxide from the air or that pandemics can be simply swatted away with new vaccine concoctions or that, among Elon Musk, children trapped within a Thai cave can be whisked to safety by a "tiny, kid-sized submarine" made from a space rocket. Surely, we will reassure ourselves, a civilization that can invent augmented reality, Hadron Colliders, and four-slice toasters should be able to replace a few bees?

Expectation is already being ladled upon projects, still in their infancy, to create genetically modified pollinators resistant to disease and chemicals or to fashion machines topped with tiny cannons that fire pollen at plants. Other scientists have turned their ingenuity to replicating the form and function of winged insects—researchers at Harvard University have devised diminutive robots that can swim before exploding out of the water into flight, using soft artificial muscles to harmlessly bounce off walls and other obstacles. Counterparts in the Netherlands have taken inspiration from the humble fruit fly, re-creating the motion of their rapid wing beats in a robot with wings made of mylar, the material used in space blankets. The Delft University of Technology's DelFly can hover, flip 360 degrees around pitch and roll axes, and accelerate to the speed of a human sprint within a few seconds.

Matěj Karásek, a researcher working on the project, says he's long been fascinated by the agility and spatial awareness of insects, even before he started working on the DelFly. "Whenever I walk outdoors and I see an insect I think 'how are they able to do this?'" he says. Karásek's robots aren't an exact substitute for a fly or bee—for one thing they have a 33-centimeter (13-inch) wingspan, making them fifty-five times the size of a fruit fly—and the conundrum of carrying large pollen payloads without losing maneuverability means they aren't quite ready to hum alongside the real thing. But there is confidence that day will arrive, drawn from the certainty many of us have that technology will eventually solve all of society's intractable ills.

Perhaps the answer will be an army of larger hexacopter-like drones, such as the fleet operated by US company Dropcopter, which autonomously pollinated an orchard of apples in New York for the first time in 2018. Or maybe a sophisticated robotic arm is the answer, which, using cameras, wheels, and artificial intelligence, can locate and hand-pollinate plants without getting tired or bored like human workers. The US Department of Agriculture is funding one such effort, which, according to one of its leading experts, Manoj Karkee, of Washington State University, promises to be a "genuine replacement for the natural pollination process" and is even "expected to be as effective or even more effective than natural pollinators like bees."

Entomologists are instinctively disdainful of any suggestion that pollinating insects could somehow be matched by technology, even on a basic logistical level. Biologist Dave Goulson points out that bees are rather adept at pollinating flowers, given they've been honing their skills for around 120 million years, and that, besides, there are around 80 million honeybee hives in the world, each stuffed with tens of thousands of bees feeding and breeding for free. "What would the cost be of replacing them with robots?" Goulson asks. "It is remarkable hubris to think that we can improve on that." To be fair to those devoted to appropriating the characteristics of insects for our use, there is widespread awe at the evolutionary brilliance of flies and bees and scant joy at the crisis that has brought us to the point where the meanderings of academic curiosity are being seized upon as possible salvation from our degenerate ways. When we consider technological solutions, we should perhaps spend less time judging the supply and more time judging the reasons why there's demand in the first place.

Still, a less abusive association with insects will have to include some new ideas. If we are to intensively farm smaller areas in order to surrender space to the wilds, the advance of vertical farming, with year-round crops stacked in warehouses and shipping containers using LED lighting and hydroponics instead of soils and pesticides, will potentially work well teamed with robotic pollinators if the original insect version demurs from the task.

Western societies may also have to grapple with the counterintuitive concept of eating insects as a way of saving them. The vast tracts of land we've turned into biodiversity deserts are in many cases not even directly feeding people—a third of all viable cropland is used to produce feed for livestock, which themselves take up a quarter of the planet's ice-free habitat. Mealworms and crickets, both excellent sources of protein that can multiply to enormous numbers in tight spaces, are a less destructive alternative to traditional Western diets and would help ease agricultural-driven pressures that blight insects, such as climate change, chemical use, and land degradation. "There are far fewer environmental problems when you eat insects. They are also delicious," says Arnold van Huis, a Dutch entomologist who has dined on twenty species of insects, his favorites being roasted termites and locusts, deep fried and served with chili.

Grinding up insects and including them in familiar foodstuffs such as bread, as bakers in the United Kingdom have done, or waffles, as has been done in Belgium, is a step toward breaking down barriers of cultural squeamishness. Inevitably, Silicon Valley has become involved, with start-ups creating artisanal cricket protein bars and mini farms to allow people to rear their own edible insects, while Western restaurants have started to dabble in insect ingredients, long used in parts of Asia and Africa, that were previously considered taboo. Livestock production "comes at a drastic cost to our environment and is why we need additional, or alternative, protein sources with lower environmental costs. Bring on the insects," says Sarah Beynon, an entomologist who has set up an insect research visitor attraction in Wales that includes the United Kingdom's first restaurant that has insects on the menu full-time.

One day, perhaps robot bees could help prop up our food supply, and a revolution in the way we eat could help slow the accelerating ruination of the world's glorious archive of life. But our measures of success in averting the insect crisis should be set a little higher than that. After all, we aren't going to witness the last insect blink out, as we will with the final northern white rhinoceros or Bengal tiger. Whatever further cruelties we inflict, there will always be insects somewhere, crawling on a windowsill plant box in Chicago, nibbling at the edge of a rice paddy in Vietnam, scurrying away from flames licking at gum trees in Australia. They have the weight of numbers and diversity, even if all that noticeably remains will be the golden retrievers of insects, the ones most comfortable around humans like cockroaches and bedbugs. The robot bees probably won't even be needed for pollination because we will strain further sinews to ramp up honeybee numbers to fill the gaping void of wild bees. Much the way 96 percent of the world's mammals are now composed of us, our cattle, and our pigs, honeybees will make up a larger and larger proportion of bee life as they are subsumed into just another, albeit important, agricultural input. At great cost, we may well find a way to muddle through.

The tragedy will be how impoverished we will become, environmentally, spiritually, morally. Bumblebees, it has been discovered, can be taught to play football, will give up sleep to care for their hive's young, and can remember good and bad experiences, hinting at a form of consciousness. The violin beetle is remarkably shaped, as the name suggests, like a violin, and side-on is almost invisibly flat. The monarch butterfly is beautiful and can taste nectar through its feet. We won't lose every single thing, but that is of scant consolation when such marvels are being ripped away. "The future is a very simplified global biota," says entomologist David Wagner. "We will have bugs but we will lose the big gaudy things. Our children will have a diminished world. That's what we are giving them."

A penurious existence, one where the marrow of life has been sucked from the bones of our surroundings, of a becalmed countryside save for the machines eking food from the remaining soils, may be one of the better scenarios facing us if the crashing of insects' tiny empires isn't heeded. The latest research shows that the loss of bees is already starting to limit the supply of key food crops, such as apples, blueberries, and cherries. Insect-eating birds are now declining not only in the featureless fields of France but even in remote parts of the Amazon rainforest. Many insect populations around the world are falling by 1 to 2 percent a year, Wagner and colleagues confirmed recently, a trend he describes as "frightening." It can, and almost certainly will, get worse. This catastrophe will plunge to some sort of nadir, although we do not appear to be close to that point yet. We're still on the downward slope, to somewhere.

*

THE HISTORICAL WEIGHT of this epoch plays on the mind of Floyd Shockley, who oversees the collection of 35 million insect specimens, kept in drawers in huge metal storage cabinets spread across five floors of the Smithsonian National Museum of Natural History, a prodigious neoclassical building resplendent with doric columns and a dome that sits on Washington, DC's National Mall, a short walk past presidential merchandise vendors and occasional protestors to the White House. When I visit Shockley's office in November 2019, one of the first things I notice is a poster of a scarab beetle on the wall. Shockley, an amiable man with a short, pointed beard, is a beetle expert. "People say 'oh my' at them because of their size, their color," he says as he waves a hand at the scarabs. "I'm more interested in the small brown things. Most of the diversity is stuff that's 5 millimeters or less, chugging along."

As we walk to see the serried ranks of insect specimens, Shockley fulminates at Americans' fetish for manicured lawns, at the celebrity of honeybees over wild bees, at the impossibility of charting all insects, never mind their population trends. The collection in this research leviathan is huge—134,000 drawers and 33,000 jars and vials filled with insects from water beetles to extinct moths to tiny thrips measuring just 0.5 millimeters (0.02 inches) and largely hidden from public view. The Smithsonian houses examples of around a third of all known insect species, and its researchers comb areas such as the canopies of South American rainforests to look for more. These trips often reinforce the mounting evidence of insect decline. Shockley is not a fan of the term "insect apocalypse," but only because the term suggests a finite, contained event rather than the ceaseless degradation now underway—less lightning bolt and more pot of water being gradually boiled. "We are at the beginning of a major extinction level event," he says. "Things are just going to get worse if humanity chooses to do nothing differently."

The scale of the current crisis seems to shrivel when you take a walk through the Smithsonian's Deep Time exhibition, situated in its famous dinosaur hall. The sprawling spectacle takes visitors on a meandering walk through Earth's history, unspooling a time line that begins with the planet's creation 4.6 billion years ago. The first land-living insects arrive at a point around 410 million years ago—a species known as *Rhyniognatha hirsti*, found pancaked in Scottish sandstone—and as other organisms spread, insects become the first animals to take flight, among the first to be able to digest plants, and are pioneers in developing camouflage from early predators. From the carboniferous age through the Mesozoic era, from around 300 million years ago, the high oxygen content and tropical conditions saw many insects balloon in size—one exhibition image of a swamp forest in Illinois shows *Meganeura*, a gargantuan dragonfly-like insect that had a wingspan of up to 71 centimeters (28 inches), just shy of that of a modern mallard duck.

The Smithsonian has punctuated this stroll through geologic time with pillars denoting the five great mass extinction events of our planet to date. Insects are a mainstay through these tombstone-marked bottlenecks, predating and then surviving dinosaurs. The most recent mass extinction 66 million years ago, seemingly triggered by a 9.7-kilometer-wide (6-mile) asteroid crashing into an area of modern-day Mexico that caused wildfires, tsunamis, acid rain, and ultimately the demise of the dinosaurs, wiped out some "finicky plant eaters" among the insects, the pillar tells us, but there was subsequent recovery and diversification as mammals, latterly humans, started to jostle for terrestrial domination. Our Pyrrhic victory at the very last gasp of Earth's history to date means, for the first time, that a single species is the primary cause of an extinction episode to impact the only known life in the universe. "These events happen every 60 to 70 million years or so, so we are due," says Shockley as we reach the end of the time line near the rotunda, where an elephant stands majestically on a patch of re-created savanna. "In just 200 years of human activity we've profoundly impacted the planet. Not even insects can react to an extinction event that fast."

One of the most unnerving aspects of the insect crisis is that we don't quite know where it will lead us. Insects can appropriately be lauded as the great survivors, the amorphous cloud of life that has passed serenely through the cataclysms of deep time to occupy every nook of the Earth. But this provides no comforting guarantees over the outcome of the present annihilation. On December 31, 2020, the final day of a year of torment, came an appropriately alarming research paper. Framing their study in the contemporary context of the world's sixth mass extinction event, two American experts in paleobiology and geology considered what the previous five mass extinctions meant for insects. It is, they conceded, difficult to determine insect abundance from the fossil record, but the evidence suggests that previous losses in insect diversity were minimal. Even the Permian extinction, the great dying of 250 million years ago that wiped out nine in ten of the planet's species, was more of a "faunal turnover" than obliteration for insects, the researchers wrote. This points to a profound conclusion—that insects are now being subjected to an existential threat unprecedented in their entire history. We cannot be confident they will bounce back like they've done before because they've *never had to endure anything like this before.* "This is not insects' sixth mass extinction—in fact, it may become their first," the researchers noted.

By flattening and poisoning our landscapes, altering the chemical composition of our atmosphere, and creating biological deserts in the pursuit of progress and aestheticism, we are conducting a high-stakes experiment with hideous risks. The protracted eons of insect history only now overlap with our brief but transformative presence on Earth. We know for certain that insects predated us, and the odds are they will survive us in some form, too. It's an arrogant presumption that we will sail unscathed through the sixth mass extinction without the diversity of insect life we are laying to waste. We need them far more than they need us. The insect crisis is, from our own self-interested point of view, a human emergency.

If you put dreams of robot bee replacements on steroids, you arrive at the sci-fi thinking, embraced as a realistic escape pod by some, of humankind relocating entirely from a trashed Earth to Mars or some other planet, terraforming its barren rockface into a new technological utopia, free of the wars and pollution and stupidity. Shockley can imagine conversations about such a far-fetched strategy taking place within the nearby halls of the US government but cannot envision, even in this extreme space-age fantasy, that we would be able to cut our ties of dependence with insects.

"Obviously anywhere we want to grow food will probably need bees," he ponders, as tourists take selfies underneath the elephant. "We may well be the first invasive species to colonize another planet. But bees will be number two."

ACKNOWLEDGMENTS

Writing a first book is a daunting enough prospect even without the added challenge of doing it during a historic pandemic, so I'm grateful and relieved I was able to finish this when so many of us, to varying degrees of trauma, have struggled.

Any writing about the environment, including the insect world, is best done having been immersed in the sights, sounds, and smells of ecosystems and I was fortunate to have undertaken some remarkable journeys for this book shortly before the shutters came down on meaningful travel.

A special thanks has to go to those who helped me navigate these places, including Cuauhtémoc Sáenz-Romero, who allowed me to traipse around the mountains of central Mexico with him looking at oyamel fir trees and monarch butterflies. I was treated with hospitality and kindness at the home of Ramirez Cruz, known locally as Don Pancho, and was saddened to learn of his death from cancer just eight months after my trip to Mexico.

Floyd Shockley was an erudite and friendly guide around the National Museum of Natural History's sprawling vaults of insects, Jay Evans assisted ably with my questions and flailing attempts to get into a beekeeping suit, and Denise Qualls was good enough to offer her insights, a bag of almonds, and a kind word after I was stung on the face in the twilight of a day spent with bees in California's Central Valley. George Hansen, a no-nonsense beekeeper, was a good mixture of pleasant company and forthright views, as was Art Shapiro on our butterfly-less yet enjoyable walk.

Attempting to write about the lives of creatures as multifaceted and yet, in many cases, obscure as insects requires plenty of expert guidance, and I'm grateful to the many entomologists and other scientists who helped with basic points of accuracy. Particular thanks go to Matt Forister, Alex Lees, Erica McAlister, Dave Goulson, Simon Potts, Alex Zomchek, Anders Pape Moller, Chris Looney, and Stefanie Christmann, among others, for their time and patience. I'm thankful that Conrad Berube was able to stand my repeated questions about what it was like being attacked by murder hornets and that Coby Schal persisted in his defense of cockroaches.

The genesis of this book has plenty to do with the instincts and hard work of Zoë Pagnamenta and her superb team, while the shape and structure of it was expertly influenced by the great Quynh Do. The latter stages of this process were adroitly overseen by Melanie Tortoroli at Norton. I'm grateful to them all for their advice and guidance.

While this book encompasses a grand sweep around the world, most of it was written in a small apartment in Brooklyn that was, outside, roiled by a rampant virus and antiracism protests and, inside, often a maelstrom due to two young children and a neurotic sausage dog. My biggest thanks and credit, therefore, need to go to my wonderful wife, Lyndal, for her love, support, and sheer endurance.

NOTES

1: AN INTRICATE DANCE

5 **"Within a few decades the world would return":** Edward O. Wilson, "The Little Things That Run the World (The Importance and Conservation of Invertebrates)," *Conservation Biology* 1, no. 4 (1987): 345.

7 **Three out of every four known animal species on Earth:** David Britton, "Why Most Animals Are Insects," Australian Museum, 2020, accessed February 25, 2021, https://australian.museum/learn/animals/insects/why-most-animals-are-insects/.

7 **"You get rid of flies? You get rid of chocolate":** Janet Fang, "Ecology: A World without Mosquitoes," *Nature* 466 (2010): 432–434, https://www.nature.com/news/2010/100721/full/466432a.html.

8 **Scientists have harnessed maggots:** K. Y. Mumcuoglu, "Clinical Applications for Maggots in Wound Care," *American Journal of Clinical Dermatology* 2, no. 4 (2001): 219.

8 **oil has been extracted from the larvae:** Qing Li et al., "From Organic Waste to Biodiesel: Black Soldier Fly, *Hermetia illucens*, Makes It Feasible," *Fuel* 90, no. 4 (2011): 1545.

8 **evolutionary theory posed by Darwin:** Dave Hone, "Moth Tongues, Orchids and Darwin—The Predictive Power of Evolution," *The Guardian*, October 2, 2013, accessed February 25, 2021, https://www.theguardian.com/science/lost-worlds/2013/oct/02/moth-tongues-orchids-darwin-evolution.

9 **Another fly, with a bright, golden abdomen:** Jennifer Welsh, "Bootylicious Fly Gets Named Beyoncé," *Live Science*, January 13, 2012, accessed February 25, 2021, https://www.livescience.com/17903-gold-butt-beyonce-fly.html.

10 **the African Matabele ants:** Erik Thomas Frank et al., "Saving the Injured: Rescue

Behavior in the Termite-Hunting Ant *Megaponera analis*," *Science Advances* 3, no. 4 (2017), accessed February 25, 2021, doi:10.1126/sciadv.1602187.

10 **Honeybees understand the concept of zero:** Scarlett R. Howard et al., "Numerical Ordering of Zero in Honey Bees," *Science* 360, no. 6393 (2018): 1124.

10 **"Certainly too little popular emphasis has been given":** Transcript of Edith Patch speech, *Bulletin of the Brooklyn Entomological Society*, February 1938, accessed February 25, 2021, https://archive.org/stream/bulletino323319371938broo/bulletino323319371938broo_djvu.txt.

11 **estimates vary from an eye-watering 30 million species:** Nigel E. Stork, "How Many Species of Insects and Other Terrestrial Arthropods Are There on Earth?," *Annual Review of Entomology* 63 (2018): 31.

11 **In 2016, Canadian scientists completed a DNA analysis:** Paul D. N. Hebert et al., "Counting Animal Species with DNA Barcodes: Canadian Insects," *Philosophical Transactions of the Royal Society B 371*, no. 1702 (2016), accessed February 25, 2021, doi.org/10.1098/rstb.2015.0333.

11 **there are around 10 quintillion:** "Numbers of Insects (Species and Individuals)," National Museum of Natural History, Smithsonian Institution, accessed February 25, 2021, https://www.si.edu/spotlight/buginfo/bugnos.

12 **hosts 3.5 trillion migrating flying insects a year:** Matt McGrath, "Trillions of High-Flying Migratory Insects Cross over UK," BBC News, December 22, 2016, accessed February 25, 2021, https://www.bbc.com/news/science-environment-38406491.

12 **scrunched them into a giant ball:** Yinon M. Bar-On, Rob Phillips, and Ron Milo, "The Biomass Distribution on Earth," *Proceedings of the National Academy of Sciences of the USA* 15, no. 25 (2018): 6506.

12 **"Today's human population is adrift in a sea of insects,":** Larry Pedigo and Marlin Rice, *Entomology and Pest Management*, 6th ed. (Long Grove, IL: Pearson College Division, 2008), 1.

12 **Bumblebees have been found at 5,500 meters:** Ian Johnston, "Bumblebees Set New Insect Record for High-Altitude Flying," *The Independent*, October 23, 2011, accessed February 25, 2021, https://www.independent.co.uk/news.

12 **a species of butterfly has an eye on its penis:** Damian Carrington, "Humanity Must Save Insects to Save Ourselves, Leading Scientist Warns," *The Guardian*, May 7, 2019, accessed February 25, 2021, https://www.theguardian.com/environment.

13 **A warning shot was fired in 2014:** Roldofo Dirzo et al., "Defaunation in the Anthropocene," *Science* 345, no. 6195 (2014): 401.

14 **treehopper was named after Lady Gaga:** Brigit Katz, "Insect with 'Wacky Fashion Sense' Named after Lady Gaga," *Smithsonian Magazine*, March 17, 2020, accessed February 25, 2021, https://www.smithsonianmag.com/smart-news/insect-wacky-fashion-sense-named-after-lady-gaga-180974435/.

14 **"Flying Ant Scenes 'Like a Horror Film'":** Emilia Bona, "Flying Ant Scenes 'Like a Horror Film' as Swarms of Insects Plague Merseyside," *Liverpool Echo*, July 12, 2020, accessed February 25, 2021, https://www.liverpoolecho.co.uk/news/liverpool-news/flying-ant-scenes-like-horror-18585600.

14 **"More Than 75 Percent Decline over 27 Years":** Caspar A. Hallmann et al., "More

Than 75 Percent Decline over 27 Years in Total Flying Insect Biomass in Protected Areas," *PLOS One* 12, no. 10 (2017), accessed February 25, 2021, doi.org/10.1371/journal.pone.0185809.

15 **"You'll miss them when they're gone":** Cover of *National Geographic*, May 2020 issue, accessed February 25, 2021, https://nationalgeographicpartners.com/2020/04/magazine-highlights-may-2020/.

15 **"Compassion for the Weevil!":** Thierry Hoquet, "Compassion pour le charançon! Vers une nouvelle philosophie de l'insecte," *Le Monde*, November 24, 2017, accessed February 25, 2021, https://www.lemonde.fr/idees/article/2017/11/24/compassion-pour-le-charancon-vers-une-nouvelle-philosophie-de-l-insecte_5219507_3232.html.

16 **"We never expected to get so many emails":** Eric Campbell, "'Insect Armageddon': Europe Reacts to Alarming Findings about Decline in Insects," ABC News, October 14, 2019, accessed February 25, 2021, https://www.abc.net.au/news/2019-10-15/insect-armageddon-europe-reacts-to-alarming-insect-decline/11593538.

19 **A paper by twenty-five researchers ominously titled:** Pedro Cardoso et al., "Scientists' Warning to Humanity on Insect Extinctions," *Biological Conservation* 242, no. 108426 (2020), accessed February 25, 2021, doi.org/10.1016/j.biocon.2020.108426.

20 **1 million species across the animal kingdom:** Jonathan Watts, "Human Society under Urgent Threat from Loss of Earth's Natural Life," *The Guardian*, May 6, 2019, accessed February 25, 2021, https://www.theguardian.com/environment/2019/may/06/human-society-under-urgent-threat-loss-earth-natural-life-un-report.

21 **"The essential, interconnected web of life":** United Nations, UN Report: "Nature's Dangerous Decline 'Unprecedented'; Species Extinction Rates 'Accelerating,'" May 6, 2019, accessed February 25, 2021, https://www.un.org/sustainabledevelopment/blog/2019/05/nature-decline-unprecedented-report/.

23 **the appalling comparison with the first research trip was clear:** Bradford C. Lister and Andrés García, "Climate-Driven Declines in Arthropod Abundance Restructure a Rainforest Food Web," *Proceedings of the National Academy of Sciences of the USA* 115, no. 44 (2018): E10397–E10406, accessed February 25, 2021, doi.org/10.1073/pnas.1722477115.

24 **A published analysis by two Australia-based scientists:** Francisco Sánchez-Bayo and Kris Wyckhuys, "Worldwide Decline of the Entomofauna: A Review of Its Drivers," *Biological Conservation* 232 (2019): 8.

26 **Across the grasslands, the number of species was cut by a third:** Sebastian Seibold et al., "Arthropod Decline in Grasslands and Forests Is Associated with Landscape-Level Drivers," *Nature* 574 (2019): 671.

2: WINNERS AND LOSERS

28 **the abundance of four species of bumblebee has plummeted:** Sydney A. Cameron et al., "Patterns of Widespread Decline in North American Bumble Bees," *Proceedings of the National Academy of Sciences of the USA* 108, no. 2 (2011): 662.

28 **Franklin's bumblebee, for instance, is only found in a narrow strip:** Robbin Thorp, "Franklin's Bumble Bee," Xerces Society, accessed February 25, 2021, https://www.xerces.org/endangered-species/species-profiles/at-risk-bumble-bees/franklins-bumble-bee.

29 **the population of the American Bumble bee, *Bombus pensylvanicus*:** Victoria J. MacPhail, Leif L. Richardson, and Shiela R. Colla, "Incorporating Citizen Science, Museum Specimens, and Field Work into the Assessment of Extinction Risk of the American Bumble bee (*Bombus pensylvanicus* De Geer 1773) in Canada," *Journal of Insect Conservation* 23 (2019): 597.

29 **"thousands of species that have just disappeared from the collection":** "Expert Warns of 'Huge Decline' in Canada's Bug Population," CTV News, October 4, 2017, accessed February 25, 2021, https://www.ctvnews.ca/canada/expert-warns-of-huge-decline-in-canada-s-bug-population-1.3618579.

29 **beetle abundance has "dropped steeply" since the mid-1970s:** Jennifer E. Harris, Nicholas L. Rodenhouse, and Richard T. Holmes, "Decline in Beetle Abundance and Diversity in an Intact Temperate Forest Linked to Climate Warming," *Biological Conservation* 240 (2019), accessed February 25, 2021, doi.org/10.1016/j.biocon.2019.108219.

30 **Ohio's butterfly population has dropped by a third:** Tyson Wepprich et al., "Butterfly Abundance Declines over 20 Years of Systematic Monitoring in Ohio, USA," *PLOS One* 17, no. 7 (2019), accessed February 25, 2021, doi.org/10.1371/journal.pone.0216270.

30 **Grasshopper numbers fell by a similar amount:** Ellen A. R. Welti et al., "Nutrient Dilution and Climate Cycles Underlie Declines in a Dominant Insect Herbivore," *Proceedings of the National Academy of Sciences of the USA* 117, no. 13 (2020): 7271.

31 **the monarch butterflies that migrate to the coast:** Emma Pelton, "Thanksgiving Count Shows Western Monarchs Need Our Help More Than Ever," Xerces Society, January 23, 2020, accessed February 25, 2021, https://xerces.org/blog/western-monarchs-need-our-help-more-than-ever.

31 **mayfly populations had slumped by more than 50 percent:** Phillip M. Stepanian et al., "Declines in an Abundant Aquatic Insect, the Burrowing Mayfly, across Major North American Waterways," *Proceedings of the National Academy of Sciences of the USA* 117, no. 6 (2020): 2987.

32 **butterflies have declined by at least 84 percent in the Netherlands:** Arco J. Van Strien et al., "Over a Century of Data Reveal More Than 80% Decline in Butterflies in the Netherlands," *Biological Conservation* 234 (2019): 116.

33 **Walter Rothschild, scion of the banking family:** Kerry Lotzof, "Walter Rothschild: A Curious Life," Natural History Museum, accessed February 25, 2021, https://www.nhm.ac.uk/discover/walter-rothschild-a-curious-life.html.

33 **The total abundance of trapped moths:** Patrick Barkham, "British Moths in Calamitous Decline, Major New Study Reveals," *The Guardian*, February 1, 2013, accessed February 25, 2021, https://www.theguardian.com/environment/2013/feb/01/british-moths-calamitous-decline.

34 **nearly half of moths tracked in the English county of Norfolk:** Richard E. Walton et al., "Nocturnal Pollinators Strongly Contribute to Pollen Transport of Wild Flowers in an Agricultural Landscape," *Biology Letters* 16 (2020), accessed February 25, 2021, doi.org/10.1098/rsbl.2019.0877.

35 **the creatures are dropping in abundance by 10 percent each decade:** Callum

J. Macgregor et al., "Moth Biomass Increases and Decreases over 50 Years in Britain," *Nature Ecology & Evolution* 3 (2019): 1645.

35 **Yet another moth study, from 2014:** Richard Fox et al., "Long-Term Changes to the Frequency of Occurrence of British Moths Are Consistent with Opposing and Synergistic Effects of Climate and Land-Use Changes," *Journal of Applied Ecology* 51, no. 4 (2014): 949–957, accessed February 25, 2021, doi.org/10.1111/1365-2664.12256.

36 **Of 353 wild bee and hoverfly species:** Damian Carrington, "Widespread Losses of Pollinating Insects Revealed across Britain," *The Guardian*, March 26, 2019, accessed February 25, 2021, https://www.theguardian.com/environment/2019/mar/26/widespread-losses-of-pollinating-insects-revealed-across-britain.

36 **labeled this neglect an "unnoticed apocalypse":** Dave Goulson et al., "Reversing the Decline of Insects," Wildlife Trusts, July 2020, accessed February 25, 2021, https://www.wildlifetrusts.org/sites/default/files/2020-07/Reversing%20the%20Decline%20of%20Insects%20FINAL%2029.06.20.pdf.

37 **half as many bee species now showing up in collecting efforts:** Yao-Hua Law, "Collectors Find Plenty of Bees but Far Fewer Species Than in the 1950s," *Science News*, January 22, 2020, accessed February 25, 2021, https://www.sciencenews.org/article/collectors-find-plenty-bees-fewer-species-than-1950s.

38 **This inertia is perhaps summed up best by a 2013 paper:** David B Lindenmayer, Maxine P. Piggott, and Brendan A. Wintle, "Counting the Books While the Library Burns: Why Conservation Monitoring Programs Need a Plan for Action," *Frontiers in Ecology and the Environment* 11, no. 10 (2013): 549–555, accessed February 25, 2021, doi.org/10.1890/120220.

39 **"the noise of their whirring wings in confined spaces":** Jeff Sparrow, "The humming of Christmas beetles was once a sign of the season. Where have they gone?," *The Guardian*, December 22, 2019, accessed February 25, 2021, https://www.theguardian.com/environment/2019/dec/23/the-humming-of-christmas-beetles-was-once-a-sign-of-the-season-where-have-they-gone.

39 **between 50 and 95 percent of the animals had lost their full litters:** Lisa Cox, "Bogong Moth Tracker Launched in Face of 'Unprecedented' Collapse in Numbers," *The Guardian*, September 17, 2019, accessed February 25, 2021, https://www.theguardian.com/environment/2019/sep/17/bogong-moth-tracker-launched-in-face-of-unprecedented-collapse-in-numbers.

40 **"The worry is, if insect populations are in decline":** Andrea Wild, "Australian Researchers Call for Help to Save Our Insects," CSIRO, December 2, 2019, accessed February 25, 2021, https://www.csiro.au/en/News/News-releases/2019/Australian-researchers-call-for-help-to-save-our-insects.

41 **"We thought it was the largest in the world":** Kate Baggaley, "World's Longest Insect Is Two Feet Long," *Popular Science*, May 6, 2016, accessed February 25, 2021, https://www.popsci.com/introducing-worlds-longest-insect/.

42 **dung beetle species in the state of Pará, in the Brazilian Amazon:** Filipe M. Franca et al., "El Niño Impacts on Human-Modified Tropical Forests: Consequences for Dung Beetle Diversity and Associated Ecological Processes," *bioTropica* 52, no. 2 (2020): 252.

42 **a reduction in the density and diversity of caterpillars:** Danielle M. Salcido, "Loss of Dominant Caterpillar Genera in a Protected Tropical Forest," *Scientific Reports* 10 (2020): 422.

43 **2019 article for the journal:** Daniel H. Janzen and Winnie Hallwachs, "Perspective: Where Might Be Many Tropical Insects?" *Biological Conservation* 233 (May 2019): 102–108, https://www.sciencedirect.com/science/article/abs/pii/S0006320719303349.

43 **"The quality of some of these papers has been relatively weak":** Raphael K. Didham et al., "Interpreting Insect Declines: Seven Challenges and a Way Forward," *Insect Conservation and Diversity* 13, no. 2 (2020): 103.

44 **an "exaggerated and unlikely narrative":** Manu E. Saunders et al., "Moving On from the Insect Apocalypse Narrative: Engaging with Evidence-Based Insect Conservation," *BioScience* 70, no. 1 (2019): 80.

44 **most data come from human-dominated locations:** Graham A. Montgomery et al., "Is the Insect Apocalypse upon us? How to Find Out," *Biological Conservation* 241 (2020), accessed February 25, 2021, doi.org/10.1016/j.biocon.2019.108327.

46 **"If you search for declines, you will find declines":** Atte Komonen et al., "Alarmist by Bad Design: Strongly Popularized Unsubstantiated Claims Undermine Credibility of Conservation Science," *Rethinking Ecology* 4 (2019): 17.

49 **"Acting with imperfect knowledge":** Matthew L. Forister, Emma M. Pelton, and Scott H. Black, "Declines in Insect Abundance and Diversity: We Know Enough to Act Now," *Conservation Science and Practice* 1, no. 8 (2019): e80.

50 **successor to the Sánchez-Bayo and Wyckhuys study:** Damian Carrington, "Insect Numbers Down 25% Since 1990, Global Study Finds," *The Guardian*, April 23, 2020, accessed February 25, 2021, https://www.theguardian.com/environment/2020/apr/23/insect-numbers-down-25-since-1990-global-study-finds.

52 **pollinator decline in North America:** "Hearing Before the Subcommittee On Horticulture and Organic Agriculture," official government ed. (Washington, DC: US Government Printing Office, 2007): 8.

54 **McClenachan dug up photos from the 1950s:** Loren McClenachan, "Documenting Loss of Large Trophy Fish from the Florida Keys with Historical Photographs," *Conservation Biology* 23, no. 3 (2009): 636.

3: "ZERO INSECT DAYS"

59 **In the more than twenty years of his car-based study:** Anders Pape Møller, "Parallel Declines in Abundance of Insects and Insectivorous Birds in Denmark over 22 Years," *Ecology and Evolution* 9, no. 11 (2019): 6581.

59 **An analysis of bird trends across Europe:** Diana E. Bowler, "Long-Term Declines of European Insectivorous Bird Populations and Potential Causes," *Conservation Biology* 33, no. 5 (2019): 1120.

60 **an estimated 12.7 million pairs of breeding birds:** Fabian Schmidt, "Insect and Bird Populations Declining Dramatically in Germany," DW, October 19, 2017, accessed February 25, 2021, https://www.dw.com/en/insect-and-bird-populations-declining-dramatically-in-germany/a-41030897.

60 **bird populations across the French countryside had slumped:** Agence France-Presse, "'Catastrophe' as France's Bird Population Collapses Due to Pesticides," *The Guardian*, March 21, 2018, accessed February 25, 2021, https://www.theguardian.com/world/2018/mar/21/catastrophe-as-frances-bird-population-collapses-due-to-pesticides.

60 **researchers acoustically tracked the northern bat:** Jens Rydell et al., "Dramatic Decline of Northern Bat *Eptesicus nilssonii* in Sweden over 30 Years," *Royal Society Open Science* 7, no. 2 (2020), accessed February 25, 2021, doi.org/10.1098/rsos.191754.

60 **Birds nesting in urban areas, meanwhile:** Nilima Marshall, "Urban Bird Populations Need Insects," *The Ecologist*, May 18, 2020, accessed February 25, 2021, https://theecologist.org/2020/may/18/urban-bird-populations-need-insects.

60 **"Insects are the cornerstone":** Gábor Seress et al., "Food Availability Limits Avian Reproduction in the City: An Experimental Study on Great Tits *Parus major*," *Journal of Animal Ecology* 89, no. 7 (July 2020): 1570–1580.

61 **impoverished food supply for whip-poor-wills:** Philina A. English, David J. Green, and Joseph J. Nocera, "Stable Isotopes from Museum Specimens May Provide Evidence of Long-Term Change in the Trophic Ecology of a Migratory Aerial Insectivore," *Frontiers in Ecology and Evolution* 6, no.14 (2018), accessed February 25, 2021, doi.org/10.3389/fevo.2018.00014.

64 **the world may face an extra 1.4 million deaths a year:** Simon G. Potts et al., "Safeguarding Pollinators and Their Values to Human Well-Being," *Nature* 540, no. 7632 (2016): 220.

66 **insect pollination improved output of the Gala:** Sean M. Webber et al., "Quantifying Crop Pollinator-Dependence and Pollination Deficits: The Effects of Experimental Scale on Yield and Quality Assessments," *Agriculture, Ecosystems & Environment* 304 (2020), accessed February 25, 2021, doi.org/10.1016/j.agee.2020.107106.

67 **"insect-derived products":** Lauren Seabrooks and Longqin Hu, "Insects: An Underrepresented Resource for the Discovery of Biologically Active Natural Products," *Acta Pharmaceutica Sinica B* 7, no. 4 (July 2017): 409–426.

67 **thanatin, a natural antibiotic produced by the spined soldier bug:** Stefan U. Vetterli et al., "Thanatin Targets the Intermembrane Protein Complex Required for Lipopolysaccharide Transport in *Escherichia coli*," *Science Advances* 4, no. 11 (2018), accessed February 25, 2021, doi.org/10.1126/sciadv.aau2634.

69 **the lowly paper wasp can grasp transitive inference:** Elizabeth A. Tibbetts et al., "Transitive Inference in *Polistes* Paper Wasps," *Biology Letters* 15, no. 5 (2019), accessed February 25, 2021, doi.org/10.1098/rsbl.2019.0015.

69 **the scuttling pests are getting "closer to invincibility":** Brian Wallheimer, "Rapid Cross-Resistance Bringing Cockroaches Closer to Invincibility," Purdue University, June 25, 2019, accessed February 25, 2021, https://www.purdue.edu/newsroom/releases/2019/Q2/rapid-cross-resistance-bringing-cockroaches-closer-to-invincibility.html.

71 **the edifying task of grinding up cockroach and locust brains:** "Cockroach Brains . . . Future Antibiotics?," University of Nottingham, September 29, 2010, accessed February 25, 2021, https://exchange.nottingham.ac.uk/blog/cockroach-brains-future-antibiotics/.

71 **a climate-controlled facility in Xichang:** Stephen Chen, "A Giant Indoor Farm in China Is Breeding 6 Billion Cockroaches a Year. Here's Why," *South China Morning Post*, April 19, 2018, accessed February 25, 2021, https://www.scmp.com/news/china/society/article/2142316/giant-indoor-farm-china-breeding-six-billion-cockroaches-year.

73 **they have learned how to steal honeydew:** Daniel A. H. Peach, "The Bizarre and Ecologically Important Hidden Lives of Mosquitoes," *The Conversation*, December 2, 2019, accessed February 25, 2021, https://theconversation.com/the-bizarre-and-ecologically-important-hidden-lives-of-mosquitoes-127599.

74 **release of 750 million modified mosquitoes:** "Florida Mosquitoes: 750 Million Genetically Modified Insects to Be Released," BBC News, August 20, 2020, accessed February 25, 2021, https://www.bbc.com/news/world-us-canada-53856776.

76 ***Anopheles quadrimaculatus*, or common marsh mosquito:** Rund Abdelfatah and Ramtin Arablouei, "'Throughline': The Mosquito's Impact on the Shaping of the U.S.," NPR, April 28, 2020, accessed February 25, 2021, https://www.npr.org/2020/04/28/846919774/throughline-the-mosquitos-impact-on-the-shaping-of-the-u-s.

4: THE PEAK OF THE PESTICIDE

79 **an impressive cave network was inhabited:** Ben Johnson, "Castleton, Peak District," Historic UK, accessed February 25, 2021, https://www.historic-uk.com/HistoryMagazine/DestinationsUK/Castleton-Peak-District/.

80 **Half of the country's ancient forest has been vanishing:** D. A. Ratcliffe, "Post-Medieval and Recent Changes in British Vegetation: The Culmination of Human Influence," *New Phytologist* 98, no. 1 (1984): 73.

81 **the number of British farms has fallen by two-thirds:** Robert A. Robinson, "Post-War Changes in Arable Farming and Biodiversity in Great Britain," *Journal of Applied Ecology* 39, no. 1 (2002): 157.

82 **80 percent of the United Kingdom's chalk grasslands:** "What's Special about Chalk Grassland?," National Trust, accessed February 25, 2021, https://www.nationaltrust.org.uk/features/whats-special-about-chalk-grassland.

83 **In his book *The Accidental Countryside*:** Amy Fleming, "Accidental Countryside: Why Nature Thrives in Unlikely Places," *The Guardian*, March 13, 2020, accessed February 25, 2021, https://www.theguardian.com/environment/2020/mar/13/accidental-countryside-why-nature-thrives-in-unlikely-places.

84 **arable fields need 10 percent weed cover:** Barbara M. Smith et al., "The Potential of Arable Weeds to Reverse Invertebrate Declines and Associated Ecosystem Services in Cereal Crops," *Frontiers in Sustainable Food Systems* 3, no. 118 (2020), accessed February 25, 2021, doi.org/10.3389/fsufs.2019.00118.

85 **meadow pipit and skylark:** "Bird Populations in French Countryside 'Collapsing,'" Phys.org, March 20, 2018.

85 **the pale giant oak aphid:** Patrick Barkham, "Ants Run Secret Farms on English Oak Trees, Photographer Discovers," *The Guardian*, January 24, 2020, accessed February 25, 2021, https://www.theguardian.com/environment/2020/jan/24/ants-run-secret-farms-on-english-oak-trees-photographer-discovers.

86 **"Wild creatures, like men":** Rachel Carson, *Silent Spring* (Boston: Houghton Mifflin, 1962).

86 **even ecosystems as vast as the Amazon:** Gregory S. Cooper et al., "Regime Shifts Occur Disproportionately Faster in Larger Ecosystems," *Nature Communications* 11, no. 1175 (2020), accessed February 25, 2021, doi.org/10.1038/s41467–020–15029-x.

86 **"increases a country's economic and food security vulnerability":** Marcelo A. Aizen et al., "Global Agricultural Productivity Is Threatened by Increasing Pollinator Dependence without a Parallel Increase in Crop Diversification," *Global Change Biology* 25, no. 10 (2019): 3516.

86 **more than a third of the world's conserved land:** Kendall R. Jones et al., "One-Third of Global Protected Land Is under Intense Human Pressure," *Science* 360, no. 6390 (2018): 788.

89 **"all just crashed and burned at the end of the year":** Adam G. Dolezal et al., "Native Habitat Mitigates Feast–Famine Conditions Faced by Honey Bees in an Agricultural Landscape," *Proceedings of the National Academy of Sciences of the USA* 116, no. 50 (2019): 25147–25155, accessed February 25, 2021, doi.org/10.1073/pnas.1912801116.

90 **the median farm size more than doubling:** James M. MacDonald and Robert A. Hoppe, "Large Family Farms Continue to Dominate U.S. Agricultural Production," USDA, March 6, 2017, accessed February 25, 2021, https://www.ers.usda.gov/amber-waves/2017/march/large-family-farms-continue-to-dominate-us-agricultural-production/.

91 **"one in a 100-year discovery":** Stephen B. Powles, "Gene Amplification Delivers Glyphosate-Resistant Weed Evolution," *Proceedings of the National Academy of Sciences of the USA* 107, no. 3 (2010): 955.

91 **glyphosate, for example, is thought to disturb bees' gut bacteria:** Erick V. S. Motta, Kasie Raymann, and Nancy A. Moran, "Glyphosate Perturbs the Gut Microbiota of Honey Bees," *Proceedings of the National Academy of Sciences of the USA* 115, no. 41 (2018): 10305–10310, accessed February 25, 2021, doi.org/10.1073/pnas.1803880115.

91 **correlation between fungicide use and the loss of bees:** Scott H. McArt et al., "Landscape Predictors of Pathogen Prevalence and Range Contractions in US Bumblebees," *Proceedings of the Royal Society B* 284, no. 1867 (2017), accessed February 25, 2021, https://doi.org/10.1098/rspb.2017.2181.

91 **fungicides can worsen outbreaks of *Nosema*:** Jeffrey S. Pettis et al., "Crop Pollination Exposes Honey Bees to Pesticides Which Alters Their Susceptibility to the Gut Pathogen *Nosema ceranae*," *PLOS One* 8, no. 7 (2013), accessed February 25, 2021, doi.org/10.1371/journal.pone.0070182.

93 **almost every single sample contained neonicotinoids:** Tao Zhang et al., "A Nationwide Survey of Urinary Concentrations of Neonicotinoid Insecticides in China," *Environment International* 132 (2019), accessed February 25, 2021, doi.org/10.1016/j.envint.2019.105114.

93 **Bayer, the maker of the chemical, to pay compensation:** Bernhard Warner, "Invasion of the 'Frankenbees': The Danger of Building a Better Bee," *The Guardian*, October 16, 2018, accessed February 25, 2021, https://www.theguardian.com/environment/2018/oct/16/frankenbees-genetically-modified-pollinators-danger-of-building-a-better-bee.

93 **Brazil is increasingly awash in agricultural chemicals:** Pedro Grigori, "Um em cada 5 agrotóxicos liberados no último ano é extremamente tóxico," *Publica*, January 16, 2020, accessed February 25, 2021, https://apublica.org/2020/01/um-em-cada-5-agrotoxicos-liberados-no-ultimo-ano-e-extremamente-toxico/.

93 **the neonicotinoid era has been a punishingly cruel one:** L. W. Pisa et al., "Effects of Neonicotinoids and Fipronil on Non-Target Invertebrates," *Environmental Science and Pollution Research* 22 (2014), accessed February 25, 2021, doi.org/10.1007/s11356-014-3471-x.

93 **a single teaspoon of imidacloprid:** "Neonicotinoids at 'Chronic Levels' in UK Rivers, Study Finds," BBC News, December 14, 2017, accessed February 25, 2021, https://www.bbc.com/news/uk-england-suffolk-42354947.

94 **only 5 percent of the chemical actually stays:** Thomas James Wood and Dave Goulson, "The Environmental Risks of Neonicotinoid Pesticides: A Review of the Evidence Post 2013," *Environmental Science and Pollution Research International* 24 (2017): 17285–325, accessed February 25, 2021, https://doi.org/10.1007/s11356-017-9240-x.

94 **the demise of butterflies, mayflies, dragonflies:** Michelle L. Hladik, Anson R. Main, and Dave Goulson, "Environmental Risks and Challenges Associated with Neonicotinoid Insecticides," *Environmental Science & Technology* 52, no. 6 (2018): 3329.

94 **clothianidin has been linked to cognitive damage:** Saija Piiroinen and Dave Goulson, "Chronic Neonicotinoid Pesticide Exposure and Parasite Stress Differentially Affects Learning in Honeybees and Bumblebees," *Proceedings of the Royal Society B* 283, no. 1828 (2016), accessed February 25, 2021, doi.org/10.1098/rspb.2016.0246.

94 **bees blighted by imidacloprid:** Daniel Kenna et al., "Pesticide Exposure Affects Flight Dynamics and Reduces Flight Endurance in Bumblebees," *Ecology and Evolution* 9, no. 10 (2019): 5637.

94 **Imidacloprid has been linked to blindness in flies:** Felipe Martelli et al., "Low doses of the neonicotinoid insecticide imidacloprid induce ROS triggering neurological and metabolic impairments in Drosophila," *Proceedings of the National Academy of Sciences of the USA* 117, no. 41 (2020): 25840.

94 **and colony losses among honeybees:** G. E. Budge et al., "Evidence for Pollinator Cost and Farming Benefits of Neonicotinoid Seed Coatings on Oilseed Rape," *Scientific Reports* 5 (2015), accessed February 25, 2021, doi.org/10.1038/srep12574.

94 **cutting the reproductive output of bumblebee queens:** Gemma L. Baron et al., "Pesticide Reduces Bumblebee Colony Initiation and Increases Probability of Population Extinction," *Nature Ecology & Evolution* 1, no. 9 (2017): 1308.

94 **neonicotinoids have been blamed:** B. A. Woodcock et al., "Country-Specific Effects of Neonicotinoid Pesticides on Honey Bees and Wild Bees," *Science* 356, no. 6345 (2017): 1393.

94 **when samples of honey were taken from around the world:** E. A. D. Mitchell et al., "A Worldwide Survey of Neonicotinoids in Honey," *Science* 358, no. 6359 (2017): 109.

95 **bees exposed to neonics during their larval stage:** Dylan B. Smith et al., "Insecticide Exposure during Brood or Early-Adult Development Reduces Brain Growth and Impairs Adult Learning in Bumblebees," *Proceedings of the Royal Society B* 287, no. 1922 (2020), accessed February 25, 2021, doi.org/10.1098/rspb.2019.2442.

96 **exposed to fatal levels of clothianidin:** D. Susan Willis Chan et al., "Assessment of Risk to Hoary Squash Bees (*Peponapis pruinosa*) and Other Ground-Nesting Bees from Systemic Insecticides in Agricultural Soil," *Scientific Reports* 9 (2019), accessed February 25, 2021, https://doi.org/10.1038/s41598-019-47805-1.

96 **migrating white-crowned sparrows:** Margaret L. Eng, Bridget J. M. Stutchbury, and Christy A. Morrissey, "A Neonicotinoid Insecticide Reduces Fueling and Delays Migration in Songbirds," *Science* 365, no. 6458 (2019): 1177.

96 **imidacloprid beyond a certain level:** Caspar A. Hallmann et al., "Declines in Insectivorous Birds Are Associated with High Neonicotinoid Concentrations," *Nature* 511 (2014): 341.

96 **pesticides and the decline of Lake Shinji:** Masumi Yamamuro et al., "Neonicotinoids Disrupt Aquatic Food Webs and Decrease Fishery Yields," *Science* 366, no. 6465 (2019): 620.

97 **forty-eight times more toxic to insect life:** Michael DiBartolomeis et al., "An Assessment of Acute Insecticide Toxicity Loading (AITL) of Chemical Pesticides Used on Agricultural Land in the United States," *PLOS One* 14, no. 8 (2019): e0220029.

97 **121 times more toxic over the past twenty years:** Margaret R. Douglas et al., "County-Level Analysis Reveals a Rapidly Shifting Landscape of Insecticide Hazard to Honey Bees (*Apis mellifera*) on US Farmland," *Scientific Reports* 10 (2020), accessed February 25, 2021, doi.org/10.1038/s41598-019-57225-w.

98 **little evidence to suggest neonicotinoids improved the harvest:** Spyridon Mourtzinis et al., "Neonicotinoid Seed Treatments of Soybean Provide Negligible Benefits to US Farmers," *Scientific Reports* 9 (2019), accessed February 25, 2021, doi.org/10.1038/s41598-019-47442-8.

99 **farms of all types across France:** Martin Lechenet et al., "Reducing Pesticide Use While Preserving Crop Productivity and Profitability on Arable Farms," *Nature Plants* 3 (2017), accessed February 25, 2021, doi.org/10.1038/nplants.2017.8.

100 **young bees fed the affected protein:** Jeffrey S. Pettis et al., "Pesticide Exposure in Honey Bees Results in Increased Levels of the Gut Pathogen *Nosema*," *Naturwissenschaften* 99, no. 2 (2012): 153.

101 **Monsanto, which is now part of Bayer:** Lee Fang, "The Playbook for Poisoning the Earth," *The Intercept*, January 18, 2020, accessed February 25, 2021, https://theintercept.com/2020/01/18/bees-insecticides-pesticides-neonicotinoids-bayer-monsanto-syngenta/.

101 **depicting people worried about pesticides as conspiracy theorists:** Bayer Crop Science, "Bayer for More TRANSPARENCY: Environmental Safety," YouTube video, 2:58, posted by Bayer Crop Science, October 30, 2018, https://www.youtube.com/watch?v=IIk0-aanjUY&feature=youtu.be.

101 **a sugar cube splashing into a cup of tea:** Bayer Crop Science, "Bayer for More TRANSPARENCY: Is Our Food SAFE?," YouTube video, 2:46, posted by Bayer Crop Science, May 3, 2018, https://www.youtube.com/watch?v=ZDlHkMTD0lY.

102 **sold $4.8 billion in highly hazardous pesticides:** Damian Carrington, "Firms Making Billions from 'Highly Hazardous' Pesticides, Analysis Finds," *The Guardian*, February 20, 2020, accessed February 25, 2021, https://www.theguardian.com/

environment/2020/feb/20/firms-making-billions-from-highly-hazardous-pesticides-analysis-finds.

104 **an extra 200 million metric tons:** Food and Agriculture Organization of the United Nations, "2050: A Third More Mouths to Feed," September 23, 2009, accessed February 25, 2021, http://www.fao.org/news/story/en/item/35571/icode/.

106 **Residential lawns are doused:** Ronda Kaysen, "One Thing You Can Do: Reduce Your Lawn," *New York Times*, April 10, 2019, accessed February 25, 2021, https://www.nytimes.com/2019/04/10/climate/climate-newsletter-lawns.html.

107 **"This whole business of keeping your lawn clipped":** Phoebe Weston, "Help Bees by Not Mowing Dandelions, Gardeners Told," *The Guardian*, February 1, 2020, accessed February 25, 2021, https://www.theguardian.com/environment/2020/feb/01/help-bees-not-mowing-dandelions-gardeners-told-aoe.

108 **light pollution now affects around a quarter of the globe's land surface:** Bernard Coetzee, "Light Pollution: The Dark Side of Keeping the Lights On," *The Conversation*, April 3, 2019, accessed February 25, 2021, https://theconversation.com/light-pollution-the-dark-side-of-keeping-the-lights-on-113489.

108 **fruit production can fall by 13 percent:** Aisling Irwin, "The Dark Side of Light: How Artificial Lighting Is Harming the Natural World," *Nature*, January 16, 2018, accessed February 25, 2021, https://www.nature.com/articles/d41586–018–00665–7.

5: IN THE TEETH OF THE CLIMATE EMERGENCY

111 **150 glaciers present in the mid-nineteenth century:** National Park Foundation, "America's Last Remaining Glaciers," accessed February 25, 2021, https://www.nationalparks.org/connect/blog/americas-last-remaining-glaciers.

112 **area covered by glaciers here has shrunk by 73 percent:** Myrna H. P. Hall and Daniel B. Farge, "Modeled Climate-Induced Glacier Change in Glacier National Park, 1850–2100," *BioScience* 53, no. 2 (2003): 131.

112 **"the best care-killing scenery on the continent":** Glacier Bear Retreat, "John Muir's Thought on Glacier National Park," accessed February 25, 2021, https://glacierbearretreat.com/john-muirs-thought-on-glacier-national-park/.

113 **The stone flies would be just another couple of obscure insect species:** J. Joseph Giersch et al., "Climate-Induced Glacier and Snow Loss Imperils Alpine Stream Insects," *Global Change Biology* 23, no. 7 (2016): 2577.

114 **only in the most scorching parts of the Sahara:** Chi Xu et al., "Future of the Human Climate Niche," *Proceedings of the National Academy of Sciences of the USA* 117, no. 21 (2020): 11350–55, accessed February 25, 2021, doi.org/10.1073/pnas.1910114117.

114 **current climate conditions of 115,000 species:** Damian Carrington, "Climate Change on Track to Cause Major Insect Wipeout, Scientists Warn," *The Guardian*, May 17, 2018, accessed February 25, 2021, https://www.theguardian.com/environment/2018/may/17/climate-change-on-track-to-cause-major-insect-wipeout-scientists-warn.

117 **Rocketing temperatures in the Arctic:** Pierre Rasmont et al., "Climatic Risk and Distribution Atlas of European Bumblebees," *BioRisk* 10 (2015): 1–236, accessed February 25, 2021, http://www.step-project.net/files/DOWNLOAD2/BR_article_4749.pdf.

117 **glowworm numbers have collapsed:** Tim Gardiner and Raphael K. Didham, "Glowing, Glowing, Gone? Monitoring Long-Term Trends in Glow-Worm Numbers in South-East England," *Insect Conservation and Diversity* 13, no. 2 (2020): 162.

117 **spectacular 80 percent drop in the abundance of mayflies:** Viktor Baranov et al., "Complex and Nonlinear Climate-Driven Changes in Freshwater Insect Communities over 42 Years," *Conservation Biology* 34, no. 5 (2020): 1241.

118 **bumblebee populations in North America:** Peter Soroye, Tim Newbold, and Jeremy Kerr, "Climate Change Contributes to Widespread Declines among Bumble Bees across Continents," *Science* 367, no. 6478 (2020): 685.

118 **"with many less bumblebees and much less diversity":** University of Ottawa, "Why Bumble Bees Are Going Extinct in Time of 'Climate Chaos,'" February 6, 2020, accessed February 25, 2021, https://media.uottawa.ca/news/why-bumble-bees-are-going-extinct-time-climate-chaos.

118 **nine new bee species had been discovered:** James B. Dorey, Michael P. Schwarz, and Mark I. Stevens, "Review of the Bee Genus *Homalictus* Cockerell (Hymenoptera: Halictidae) from Fiji with Description of Nine New Species," *ZooTaxa* 4674, no. 1 (2019), accessed February 25, 2021, doi.org/10.11646/zootaxa.4674.1.1.

119 **"most wonderful case of fertilisation":** Charles Darwin to J. D. Hooker, April 7, 1874, ed. Darwin Correspondence Project, *The Correspondence of Charles Darwin*, vol. 22 (Cambridge, UK: Cambridge University Press), accessed February 25, 2021, http://cudl.lib.cam.ac.uk/view/MS-DAR-00095-00321/1.

120 **moths and butterflies are emerging from their cocoons:** James R. Bell et al., "Spatial and Habitat Variation in Aphid, Butterfly, Moth and Bird Phenologies over the Last Half Century," *Global Change Biology* 25, no. 6 (2019): 1982.

120 **springtime conditions that trigger insect activity:** Angela Fritz, "Spring Is Running 20 Days Early. It's Exactly What We Expect, but It's Not Good," *Washington Post*, February 27, 2018, accessed February 25, 2021, https://www.washingtonpost.com/news/capital-weather-gang/wp/2018/02/27/spring-is-running-20-days-early-its-exactly-what-we-expect-but-its-not-good/.

121 **Richard Fox, associate director at the Butterfly Conservation charity:** Richard Fox, Twitter, April 24, 2020, accessed February 25, 2021, https://twitter.com/RichardFoxBC/status/1253723902007824384.

121 **"Don't write a butterfly book":** Patrick Barkham, "UK Butterfly Season Off to Unusually Early Start after Sunniest of Springs," *The Guardian*, June 6, 2020, accessed February 25, 2021, https://www.theguardian.com/environment/2020/jun/06/uk-butterfly-season-off-to-unusually-early-start-after-sunniest-of-springs.

123 **a different scent was given off by plants:** Coline Jaworski, Benoit Geslin, and Catherine Fernandez, "Climate Change: Bees Are Disorientated by Flowers' Changing Scents," *The Conversation*, June 26, 2019, accessed February 25, 2021, https://theconversation.com/climate-change-bees-are-disorientated-by-flowers-changing-scents-119256.

124 **yields of the three most important grain crops:** Curtis A. Deutsch et al., "Increase in Crop Losses to Insect Pests in a Warming Climate," *Science* 361, no. 6405 (2018): 916.

124 **populations of houseflies more than doubling:** Dave Goulson et al., "Predicting

Calyptrate Fly Populations from the Weather, and Probable Consequences of Climate Change," *Journal of Applied Ecology* 42, no. 5 (2005): 795.

124 **"There's not going to be a 'somebody else's problem'":** Vicky Stein, "How Climate Change Will Put Billions More at Risk of Mosquito-Borne Diseases," PBS, March 28, 2019, accessed February 25, 2021, https://www.pbs.org/newshour/science/how-climate-change-will-put-billions-more-at-risk-of-mosquito-borne-diseases.

125 **"Hordes of mosquitoes suffocated cattle":** Gordon Patterson, *The Mosquito Wars* (Gainesville, FL: University Press of Florida, 2004): foreword.

127 **attracted the attention of the *New York Times*:** Mike Baker, "'Murder Hornets' in the U.S.: The Rush to Stop the Asian Giant Hornet," *New York Times*, May 2, 2020, accessed February 25, 2021, https://www.nytimes.com/2020/05/02/us/asian-giant-hornet-washington.html.

127 **"Murder hornets. Sure thing, 2020.":** Patton Oswalt, Twitter, May 2, 2020, accessed February 25, 2021, https://twitter.com/pattonoswalt/status/1256634924997607424.

127 **"My colleagues in Japan, China and Korea":** Jeanette Marantos, "Panicked over 'Murder Hornets,' People Kill Bees We Need," *Los Angeles Times*, May 8, 2020, accessed February 25, 2021, https://www.latimes.com/lifestyle/story/2020–05–08/panicked-over-murder-hornets-people-are-killing-the-native-bees-we-desperately-need.

129 **at least twenty-eight people perished in Shaanxi:** Chris Luo, "Wave of Hornet Attacks Kills 28 in Southern Shaanxi," *South China Morning Post*, September 26, 2013, accessed February 25, 2021, https://www.scmp.com/news/china-insider/article/1318293/wave-hornet-attacks-kills-28-southern-shaanxi.

130 **the Schmidt pain index:** Natural History Museum, "The Schmidt sting pain index," accessed February 25, 2021, https://www.nhm.ac.uk/scroller-schmidt-painscale/#intro.

130 **1950s alien invasion–style illustration of hornets:** Entomological Society of British Columbia, "Giant Alien Hornet Invasion!," poster, accessed February 25, 2021, http://entsocbc.ca/wp-content/uploads/2019/10/Asian-Giant-Hornet-poster-2019.pdf.

132 **Sydney was enveloped in smoke:** Melissa Davey, "NSW bushfires: Doctors Sound Alarm over 'Disastrous' Impact of Smoke on Air Pollution," *The Guardian*, December 10, 2019, accessed February 25, 2021, https://www.theguardian.com/environment/2019/dec/10/nsw-bushfires-doctors-sound-alarm-over-disastrous-impact-of-smoke-on-air-pollution.

135 **An assessment following the Black Summer:** Pallab Ghosh, "Climate Change Boosted Australia Bushfire Risk by at Least 30%," BBC News, March 4, 2020, accessed February 25, 2021, https://www.bbc.com/news/science-environment-51742646.

6: THE LABOR OF HONEYBEES

138 **the fertile soils of the valley:** "California's Central Valley," United State Geological Survey, accessed February 25, 2021, https://ca.water.usgs.gov/projects/central-valley/about-central-valley.html.

139 **California produces 80 percent of the world's almonds:** Robert Rodriguez, "Almond Acreage in California Grows to Record Total," *The Fresno Bee*, April 26, 2018,

accessed February 25, 2021, https://www.fresnobee.com/news/business/agriculture/article209894464.html.

141 **agricultural production dependent on pollination has increased 300 percent:** "Pollinators Vital to Our Food Supply under Threat," Food and Agriculture Organization of the United Nations, February 26, 2016, accessed February 25, 2021, http://www.fao.org/news/story/en/item/384726/icode/.

142 **survey of seventeen European Union members:** Daniel Cressey, "EU States Lose Up to One-Third of Honeybees per Year," *Nature*, April 9, 2014, accessed February 25, 2021, https://www.nature.com/news/eu-states-lose-up-to-one-third-of-honeybees-per-year-1.15016.

142 **"If we lose the bees, we lose fruits, vegetables, even grains":** Ivy Scott, "French Honey at Risk as Dying Bees Put Industry in Danger," France 24, June 27, 2019, accessed February 25, 2021, https://www.france24.com/en/20190627-french-honey-bees-climate-change-pesticides-farming.

142 **Between 1985 and 2008, there was a 54 percent decline:** University of Reading, "Sustainable Pollination Services for UK Crops," accessed February 25, 2021, https://www.reading.ac.uk/web/files/food-security/cfs_case_studies_-_sustainable_pollination_services.pdf.

144 ***Time* magazine featured a solitary honeybee:** "A World Without Bees," cover of *Time*, August 19, 2013, accessed February 25, 2021, http://content.time.com/time/covers/0,16641,20130819,00.html.

145 **high-fructose corn syrup containing imidacloprid:** Harvard University, "Use of Common Pesticide Linked to Bee Colony Collapse," April 5, 2012, accessed February 25, 2021, https://www.hsph.harvard.edu/news/press-releases/colony-collapse-disorder-pesticide/.

145 **"more like having a mosquito land on you":** Peter Hess, "Bee Collapse: The *Varroa* Mite Is More Destructive Than Scientists Ever Knew," *Inverse*, January 18, 2019, accessed February 25, 2021, https://www.inverse.com/article/52529-scientists-finally-understand-why-varroa-mites-kill-bees.

146 **nearly 40 percent of managed honeybee colonies were lost:** Susie Neilson, "More Bad Buzz for Bees: Record Number of Honeybee Colonies Died Last Winter," NPR, June 19, 2019, accessed February 25, 2021, https://www.npr.org/sections/thesalt/2019/06/19/733761393/more-bad-buzz-for-bees-record-numbers-of-honey-bee-colonies-died-last-winter.

152 **"Since they are recognized and appreciated by most people":** University of Hawaii, "To Bee, or Not to Bee, a Question for Almond Growers," February 28, 2020, accessed February 25, 2021, https://www.hawaii.edu/news/2020/02/28/to-bee-or-not-to-bee/.

156 **deformed wing virus has become much more genetically diverse:** Eugene V. Ryabov et al., "Dynamic Evolution in the Key Honey Bee Pathogen Deformed Wing Virus: Novel Insights into Virulence and Competition Using Reverse Genetics," *PLOS Biology* 17, no. 10 (2019): e3000502.

158 **"long-term international or national monitoring":** S. G. Potts, V. L. Imperatriz-Fonseca, and H. T. Ngo, eds., *The Assessment Report of the Intergovernmental Science-Policy*

Platform on Biodiversity and Ecosystem Services on Pollinators, Pollination and Food Production (Bonn, Germany: IPBES, 2016).

159 **fourteen bee species in New England have declined:** Minna E. Mathiasson and Sandra M. Rehan, "Status Changes in the Wild Bees of North-Eastern North America over 125 Years Revealed through Museum Specimens," *Insect Conservation and Diversity* 12, no. 4 (2019): 278.

159 **In a review of more than 4,000 native bee species:** Kelsey Kopec and Lori Ann Burd, "Pollinators in Peril," Center for Biological Diversity, March 1, 2017, accessed February 25, 2021, https://www.biologicaldiversity.org/campaigns/native_pollinators/pdfs/Pollinators_in_Peril.pdf.

162 **highest metabolic rates recorded in any organism:** Dave Goulson, "Bumblebees," in *Silent Summer: The State of Wildlife in Britain and Ireland*, ed. Norman Maclean (Cambridge, UK: Cambridge University Press, 2010), 416.

162 **a third of British wild bees and hoverflies are in decline:** Helen Briggs, "Bees: Many British Pollinating Insects in Decline, Study Shows," BBC News, March 26, 2019, accessed February 25, 2021, https://www.bbc.com/news/science-environment-47698294.

163 **honeybees are transferring diseases to wild bees:** Samantha A. Alger et al., "RNA Virus Spillover from Managed Honeybees (*Apis mellifera*) to Wild Bumblebees (*Bombus* spp.)," *PLOS One* 14, no. 6 (2019): e0217822.

164 ***Varroa destructor* can "nimbly climb" from flowers:** David T. Peck et al., "*Varroa destructor* Mites Can Nimbly Climb from Flowers onto Foraging Honey Bees," *PLOS One* 11, no. 12 (2016): e0167798.

164 **"It's quite hip at the moment":** Palko Karasz and Christopher F. Schuetze, "Bees Swarm Berlin, Where Beekeeping Is Booming," *New York Times*, August 11, 2019, accessed February 25, 2021, https://www.nytimes.com/2019/08/11/world/europe/berlin-bees-swarm.html.

7: A MONARCH'S JOURNEY

168 **butterflies were placed in miniature flight simulators:** Henrik Mouritsen and Barrie J. Frost, "Virtual Migration in Tethered Flying Monarch Butterflies Reveals Their Orientation Mechanisms," *Proceedings of the National Academy of Sciences of the USA* 99, no. 15 (2002): 10162.

168 **placing small, circular tags on more than a million:** Elizabeth Pennisi, "Mysterious Monarch Migrations May Be Triggered by the Angle of the Sun," *Science*, December 18, 2019, accessed February 25, 2021, https://www.sciencemag.org/news/2019/12/mysterious-monarch-migrations-may-be-triggered-angle-sun.

170 **"If things stay the same, western monarchs probably won't be around":** Eric Sorensen, "Monarch Butterflies Disappearing from Western North America," Washington State University news, September 7, 2017, accessed February 25, 2021, https://news.wsu.edu/2017/09/07/monarch-butterflies-disappearing/.

170 **since 1990, close to 1 billion monarch butterflies had vanished:** Darryl Fears,

"The Monarch Massacre: Nearly a Billion Butterflies Have Vanished," *Washington Post*, February 9, 2015, accessed February 25, 2021, https://www.washingtonpost.com/news/energy-environment/wp/2015/02/09/the-monarch-massacre-nearly-a-billion-butterflies-have-vanished/.

171 **A 2012 research paper coauthored by Sáenz-Romero:** Cuauhtémoc Sáenz-Romero et al., "*Abies religiosa* Habitat Prediction in Climatic Change Scenarios and Implications for Monarch Butterfly Conservation in Mexico," *Forest Ecology and Management* 275 (2012): 98.

178 **nearly half of the city-state's native butterfly species:** Meryl Theng et al., "A Comprehensive Assessment of Diversity Loss in a Well-Documented Tropical Insect Fauna: Almost Half of Singapore's Butterfly Species Extirpated in 160 Years," *Biological Conservation* 242 (2020), accessed February 25, 2021, doi.org/10.1016/j.biocon.2019.108401.

179 **40 percent of common butterfly species declined in number:** "Drastic Decline in Japan's Butterfly Population; Other Wildlife Also Feared Endangered," *The Mainichi*, November 17, 2019, accessed February 25, 2021, https://mainichi.jp/english/articles/20191116/p2a/00m/0na/023000c.

181 **Populations of grassland butterfly species plummeted:** "Populations of Grassland Butterflies Decline Almost 50 % over Two Decades," European Environment Agency, July 17, 2013, accessed February 25, 2021, https://www.eea.europa.eu/highlights/populations-of-grassland-butterflies-decline.

182 **a Bath white butterfly, a rare visitor to the United Kingdom:** "*Pontia daplidice* (circa 1702) [OUMNH]," UK Butterflies, accessed February 25, 2021, https://www.ukbutterflies.co.uk/album_photo.php?id=14265.

184 **Since 1976, habitat-specialist butterflies:** Martin S. Warren et al., "The Decline of Butterflies in Europe: Problems, Significance, and Possible Solutions," *Proceedings of the National Academy of Sciences of the USA 118*, no. 2 (2020): e2002551117.

184 **the "state of the nation" for butterflies:** "The State of Britain's Butterflies," Butterfly Conversation, 2015, accessed February 25, 2021, https://butterfly-conservation.org/butterflies/the-state-of-britains-butterflies.

185 **Duke of Burgundy butterfly shrank by 84 percent:** Patrick Barkham, "The Butterfly Effect: What One Species' Miraculous Comeback Can Teach Us," *The Guardian*, May 27, 2019, accessed February 25, 2021, https://www.theguardian.com/environment/2019/may/27/butterfly-miraculous-comeback-save-planet-duke-burgundy.

185 **A major 2013 study of 337 common UK moth species:** "The State of Britain's Larger Moths 2013," Butterfly Conservation, 2013, accessed February 25, 2021, https://butterfly-conservation.org/sites/default/files/1state-of-britains-larger-moths-2013-report.pdf.

185 **butterfly species declined by 58 percent on farmed land:** Andre S. Gilburn et al., "Are Neonicotinoid Insecticides Driving Declines of Widespread Butterflies?," *PeerJ* 3 (2015), accessed February 25, 2021, doi.org/10.7717/peerj.1402.

193 **"We should care about monarchs like we care about the Mona Lisa":** Elizabeth Howard, "Farewell, Dr. Lincoln Brower," Journey North, July 23, 2018, accessed February 25, 2021, https://journeynorth.org/monarchs/news/spring-2018/071718-dr-lincoln-brower.

8: THE INACTION PLAN

194 **The petitioners called for 30 percent of farmland:** Kate Connolly, "Bavaria Campaigners Abuzz as Bees Petition Forces Farming Changes," *The Guardian*, February 14, 2019, accessed February 25, 2021, https://www.theguardian.com/world/2019/feb/14/bavaria-campaigners-abuzz-as-bees-petition-forces-farming-changes.

195 **"suffice to crack the atom, to command the tides":** Aldo Leopold, *The River of the Mother of God and Other Essays by Aldo Leopold*, ed. Susan L. Flader and J. Baird Callicott (Madison, WI: The University of Wisconsin Press, 1991), 254.

198 **topsoil is lost through erosion worldwide:** Food and Agriculture Organization of the United Nations, "Status of the World's Soil Resources," 2015, accessed February 25, 2021, http://www.fao.org/3/i5228e/I5228E.pdf.

199 **as the title of Tree's book on Knepp alludes:** Isabella Tree, *Wilding* (London: Picador, 2018).

200 **we must protect everything, from the cathedrals of rainforests:** Pedro Cardoso et al., "Scientists' Warning to Humanity on Insect Extinctions," *Biological Conservation* 242 (2020), accessed February 25, 2021, doi.org/10.1016/j.biocon.2020.108426.

201 **there is a far more farmer-friendly alternative:** Stefanie Christmann et al., "Farming with Alternative Pollinators Increases Yields and Incomes of Cucumber and Sour Cherry," *Agronomy for Sustainable Development* 37, no. 24 (2017), accessed February 25, 2021, doi.org/10.1007/s13593-017-0433-y.

202 **Buglife, a British insect conservation group:** Buglife, "B-Lines," accessed February 25, 2021, https://www.buglife.org.uk/our-work/b-lines/.

204 **plume of oil spreading into the waterway:** Newtown Creek Alliance, "Greenpoint Oil Spill," accessed February 25, 2021, http://www.newtowncreekalliance.org/greenpoint-oil-spill.

206 **Utrecht is transforming bus shelters into bee sanctuaries:** Michiel de Gooijer, "This Dutch City Has Transformed Its Bus Stops into Bee Stops," *EcoWatch*, July 8, 2019, accessed February 25, 2021, https://www.ecowatch.com/dutch-city-bus-stops-into-bee-stops-2639127437.html.

209 **There's even a specimen of an ancient robber fly:** Erica McAlister, "Celebrating Robber Flies—Big, Beautiful Venomous Assassins," Natural History Museum, April 30, 2018, accessed February 25, 2021, https://naturalhistorymuseum.blog/2018/04/30/celebrating-robber-flies-big-beautiful-venomous-assassinscurator-of-diptera/.

9: A HUMAN EMERGENCY

213 **Dropcopter, which autonomously pollinated an orchard:** Christina Herrick, "New York Apple Orchard Claims World First in Pollination by Drone," *Growing Produce*, June 12, 2018, accessed February 25, 2021, https://www.growingproduce.com/fruits/apples-pears/new-york-apple-orchard-claims-world-first-in-pollination-by-drone/.

213 **"genuine replacement for the natural pollination process":** Scott Weybright, "Robotic Crop Pollination Awarded $1 Million Grant," Washington State University,

June 19, 2020, accessed February 25, 2021, https://news.wsu.edu/2020/06/19/robotic-crop-pollination-goal-new-1-million-grant/.

214 **"What would the cost be of replacing them with robots?":** Dave Goulson, "Are Robotic Bees the Future?" University of Sussex blog, accessed February 25, 2021, http://www.sussex.ac.uk/lifesci/goulsonlab/blog/robotic-bees.

215 **first restaurant that has insects on the menu full-time:** Sarah Benyon, "Bug Burgers, Anyone? Why We're Opening the UK's First Insect Restaurant," *The Conversation*, October 22, 2015, accessed February 25, 2021, https://theconversation.com/bug-burgers-anyone-why-were-opening-the-uks-first-insect-restaurant-49078.

215 **96 percent of the world's mammals:** Damian Carrington, "Humans Just 0.01% of All Life but Have Destroyed 83% of Wild Mammals—Study," *The Guardian*, May 21, 2018, accessed February 25, 2021, https://www.theguardian.com/environment/2018/may/21/human-race-just-001-of-all-life-but-has-destroyed-over-80-of-wild-mammals-study.

216 **the loss of bees is already starting to limit:** J. R Reilly et al., "Crop Production in the USA Is Frequently Limited by a Lack of Pollinators," *Proceedings of the Royal Society B: Biological Sciences* 287, no. 1931 (2020), accessed February 25, 2021, doi.org/10.1098/rspb.2020.0922.

216 **Insect-eating birds are now declining:** Daniel Grossman, "Nine Insect-Eating Bird Species in Amazon in Sharp Decline, Scientists Find," *The Guardian*, October 26, 2020, accessed February 25, 2021, https://www.theguardian.com/environment/2020/oct/26/nine-insect-eating-bird-species-in-amazon-in-sharp-decline-scientists-find.

216 **Many insect populations around the world are falling:** David L. Wagner et al., "Insect Decline in the Anthropocene: Death by a Thousand Cuts," *Proceedings of the National Academy of Sciences of the USA* 118, no. 2 (2021): e2023989118.

218 **a species known as *Rhyniognatha hirsti*:** Paul Rincon, "Oldest Insect Delights Experts," BBC News, February 11, 2004, accessed February 25, 2021, http://news.bbc.co.uk/2/hi/science/nature/3478915.stm.

218 ***Meganeura*, a gargantuan dragonfly-like insect:** Ker Than, "Giant Bugs Eaten Out of Existence by First Birds?," *National Geographic*, June 5, 2012, accessed February 25, 2021, https://www.nationalgeographic.com/animals/article/120601-insects-birds-giant-prehistoric-clapham-proceedings-science-bugs.

219 **December 31, 2020, the final day of a year of torment:** Sandra R. Schachat and Conrad C. Labandeira, "Are Insects Heading Toward Their First Mass Extinction? Distinguishing Turnover from Crises in Their Fossil Record," *Annals of the Entomological Society of America* (2020), accessed February 25, 2021, doi.org/10.1093/aesa/saaa042.

220 **tourists take selfies underneath the elephant:** "African Bush Elephant," National Museum of Natural History, Smithsonian Institution, accessed February 25, 2021, https://naturalhistory.si.edu/exhibits/african-bush-elephant.

INDEX